Praise for *Vax-Unvax*

"Millions of people—myself included—initially believed the COVID-19 vaccine disaster to be a one-off, the result of a novel, rapidly evolving virus combined with a rushed therapeutic packaged in an experimental delivery system. Today I laugh at such naïveté. In *Vax-Unvax*, Kennedy and Hooker shine a blinding light on the appalling lack of research and blatant propaganda behind the entire inflated and ever-expanding childhood vaccine schedule. The authors' painstaking investigation and rigorous analyses are rivaled only by their bravery in exposing the depth and breadth of the lies we've been told. As a physician who never dreamed of questioning the safety and efficacy claims of routine immunizations and who believed he was protecting his patients and his own children by endorsing them, I am humbled and enraged. Our government, the media, and the powerful and rapacious pharmaceutical industrial complex have deceived, endangered, and gaslit the public for far too long. I hope this explosive and important book finds a worldwide audience and becomes a staple in every pediatrician's and parent's library."

—Dr. Pierre Kory, author of *The War on Ivermectin*, cofounder of the Front Line COVID-19 Critical Care Alliance, cofounder of the Leading Edge Tele-Health Clinic

"As the threat of fatal infectious diseases in childhood has dissipated, there has been a lagged increase in the intensity of vaccination for rare illnesses. Modern treatment and supportive care have taken much of the concern out of the infections on the childhood vaccine schedule. However, with the rise of hyper-vaccination, Kennedy and Hooker uncover a concurrent explosion of childhood allergic, immune, and neuropsychiatric illnesses. Massive systemic perturbation of the immune system with indiscriminate immunization has come at a sobering cost. Read *Vax-Unvax* carefully and keep it close at hand—this dawn of a new age in public health will be tumultuous for years to come."

—Dr. Peter McCullough, author of *The Courage to Face COVID-19*

"In this book, Kennedy and Hooker provide the complete, definitive demolition of the mythos and propaganda that tells the public vaccines improve the health of children. Not a plank of this false house remains by the end of this book. The lies are dismantled in cool, clear language void of bombast, which allows the facts, figures, and data to shine through, to a devastating conclusion. This is the book you can hand to people who are still in trance states about vaccines."

—Celia Farber, journalist and author of *Serious Adverse Events*

"'The Science' is finally here in one place. If you want to follow the science related to vaccines and health problems, this is a must read. With over one hundred references, the actual harm being caused by vaccines is exposed. Parents, don't listen to an authority figure without doing your own research. This book is required reading for every informed parent."

—**Paul Thomas, MD, author of** *The Vaccine-Friendly Plan*
and *The Addiction Spectrum*; **founder and host of** *With the Wind: Science*
Revealed; **cofounder KidsFirst4Ever.com**

"This clear, compelling, timely book lays to rest most myths about the 'science' and safety of many existing vaccines and exposes shoddy testing, shocking damage to health, and corrupt business practices. An important follow-up to Kennedy's *The Real Anthony Fauci*."

—**Naomi Wolf, bestselling author of** *The Beauty Myth*
and *The Bodies of Others*

"While the CDC continues to refuse to do the type of vaccinated versus unvaccinated study that parents have long been demanding, independent researchers have forged ahead, and the results are now quite clear: unvaccinated children are healthier. In *Vax-Unvax: Let the Science Speak*, Robert F. Kennedy Jr. and Dr. Brian Hooker review those data as well as many additional studies comparing health outcomes between those who did or did not receive individual vaccines. This is an essential resource for the serious researcher and a valuable guide for anyone wishing to exercise truly informed consent. The graphs helpfully included with the discussion of each study covered speak for themselves. It is past time for the 'public health' establishment to stop deceiving the public with their proclamations of official dogma about the ostensible safety of these pharmaceutical products."

—**Jeremy R. Hammond, independent journalist**
and author of *The War on Informed Consent*

"In 1999, I gave a vaccine that caused such a bad side effect that it altered the trajectory of my patient's life. Thus began my journey into vaccine safety research. This book is a great compilation of scientific studies you never heard about on the news. With easy-to-understand graphics and explanations of statistics, you can analyze data from clinicians and researchers from around the world. You may find yourself doubting the simplistic 'safe and effective' doctrine repeated by health authorities. You will find yourself more empowered to make vaccine decisions for your child."

—**Elizabeth Mumper, MD, IFMCP, president & CEO,**
Rimland Center for Integrative Medicine

"RFK Jr. and Dr. Hooker present the science that supports what I have personally witnessed in my twenty-five years as a pediatrician—unvaccinated children are healthier and have fewer chronic medical problems compared to vaccinated kids. Today's parents, and a growing number of my colleagues, are now coming to recognize this grand irony in our modern pediatric health care system."

—Dr. Bob Sears, author of *The Vaccine Book* and host of TheVaccineConversation.com podcast

"Bobby Kennedy and Brian Hooker are tireless heroes on the front line of a great battle to protect our health freedom. Biomedical studies are being falsified and the masses are being deceived by health authorities, the vaccine industry, and a complicit media. Read this book and let the true science speak!"

—Neil Z. Miller, author of *Miller's Review of Critical Vaccine Studies*

"When I met Dr. Hooker on August 29, 2014, he was more than a decade into his relentless efforts through the Freedom of Information Act (FOIA) to reveal not only the studies the CDC refused to do but the plague of corruption surrounding CDC whistleblower William Thompson's confession regarding the censorship and fraud in the MMR vaccine trial by the CDC to cover up the fraud in the in clinical literature surrounding all vaccines. Stunned by the blatant corruption in Dr. Hooker's presentation that day, Kent Heckenlively and I, in collaboration with Robert F. Kennedy Jr., began to write *Plague of Corruption*. Throughout 2019, Dr. Hooker and Kennedy collaborated to reveal much of the CDC, FDA, and NIH corruption in the thirty-four page foreword to *Plague of Corruption* that revealed manipulation of hundreds of basic research studies showing dangers of xenotransplantation, microbial contamination, and environmental toxins including mercury, aluminum, PEG, and the corruption of the agencies tasked to protect public health.

Vax-Unvax is the result of their heroic effort to reveal the censored science and truth behind the role of a failure by these agencies tasked to conduct safety studies in three decades of liability-free vaccines and the resultant explosion of chronic disease and disability facing our world today. Given the massive push to vaccinate a new generation, this book is a must read supporting a moratorium on inoculations in favor of oral and mucosal immunization strategies."

—Judy A. Mikovits, PhD, author of *Plague of Corruption*

"In *Vax-Unvax: Let the Science Speak*, authors Robert F. Kennedy Jr. and Brian Hooker have provided a deep dive into vaccine safety by looking at published data from independent researchers. The resulting review of vaccinated and unvaccinated children's health outcomes clearly shows the damage, much of it neurological, that blind adherence to the CDC's current vaccine schedule can induce. The studies highlighted are those that both the FDA and CDC have routinely refused, and continue to refuse, to do themselves. The official reason for not providing such studies is based on the erroneous notion that conducting vaccinated-unvaccinated evaluations would somehow be 'unethical.' Kennedy and Hooker demolish this argument and then proceed to review the safety of different vaccines, many that contain adjuvant aluminum or Thimerosal, the latter an ethyl mercury compound. Overall, for those trying to understand the often-confusing claims and counterclaims, particularly lay people, the book provides some badly needed clarity.

The book also considers the current COVID-19 mRNA vaccines in the context of the older childhood vaccine platforms. This is a particularly timely contribution in that the COVID-19 pandemic and resulting vaccine mandates have paradoxically served to make more people question the official 'safe and effective' mantra that tends to surround all vaccines: If the conventional vaccine platforms using compounds with aluminum and mercury are not safe, then why should anyone trust the completely new and largely experimental vaccine platforms developed for COVID (and soon numerous vaccines)? For parents it can be both confusing and frightening to consider the pros and cons of vaccines against childhood diseases: what if they make the wrong decision, in either direction, and their child is harmed?

While many in the medical profession may not like the book because it exposes the outright deception of the pharmaceutical industry, the CDC, and the FDA, I strongly believe that many parents, or parents-to-be, will be grateful for the information it contains. Simply put, at the end of the day the ability to consider all aspects of vaccine safety in order to make an informed choice for one's children, or oneself, is an absolutely critical aspect of real health freedom. In turn, health freedom is intimately tied to the concept of 'security of the person,' perhaps the most fundamental of natural rights.

Kennedy and Hooker should be commended for tackling this crucial issue in order to bring clarity to the mass of 'dis-' and 'mis-' information peddled by the health establishment and the mainstream media. If indeed the 'truth can set you free,' then this book is a huge step in the right direction."

—Christopher Shaw, neuroscientist and professor of ophthalmology at University of British Columbia, author of *Dispatches from the Vaccine Wars*

"If there is only one book you read in your entire life, let it be this one! If you want the science to speak . . . then have the courage to look at the actual science, data, and truth found in the pages of *Vax-Unvax*. Arm yourself with the information that puts the power back into the hands of parents where it belongs, not in the hands of corrupt Pharma, captured government officials, and incentivized doctors spouting catchphrases with little to no evidence to back it up. RFK Jr. and Dr. Brian Hooker are the bold voices of truth, presenting evidence that cannot be disputed. The main purpose of a parent is to love your child and keep them safe. If you have not had the courage before, I boldly implore you to find the courage now, and educate before you vaccinate!"

—Leigh-Allyn Baker, actress, producer, and star of the global hit *Good Luck Charlie*

Vax-Unvax

LET THE
SCIENCE SPEAK

Robert F. Kennedy Jr.
NEW YORK TIMES BESTSELLING AUTHOR

•••••••••••••••••••••••••• AND ••••••••••••••••••••••••••

Brian Hooker, PhD

Foreword by Del Bigtree

Skyhorse Publishing

Children's
Health Defense

Skyhorse Publishing books may be purchased in bulk at special discounts for
sales promotion, corporate gifts, fund-raising, or educational purposes. Special
editions can also be created to specifications. For details, contact the Special Sales
Department, Skyhorse Publishing, 307 West 36th Street, 11th Floor, New York,
NY 10018 or info@skyhorsepublishing.com

Skyhorse® and Skyhorse Publishing® are registered trademarks of Skyhorse
Publishing, Inc.®, a Delaware corporation.

Visit our website at www.skyhorsepublishing.com.
Please follow our publisher Tony Lyons on Instagram @tonylyonsisuncertain

10 9 8 7 6 5 4 3 2 1

Library of Congress Cataloging-in-Publication Data is available on file.

Hardcover ISBN: 978-1-5107-6696-9
eBook ISBN: 978-1-5107-7-6697-6

Cover design by Brian Peterson

Printed in the United States of America

Contents

Dedication

From as early as 2002, parents and other advocates have been asking the United States government for a study on health outcomes in fully vaccinated versus unvaccinated children. This book, full of vax-unvax studies, many of which were unintentional but paramount, is a gift to those people and to all who value the truth.

For years, advocates thought they could make vax-unvax studies happen if they got into the right places, became appointed to scientific panels, and held special meetings with people in power. After their hard work attained those exact goals, the advocates realized that public health officials would never allow such an undertaking. The very agencies covering up wrongdoing against children would never conduct such a comprehensive study, the results of which would be damning to the vaccine orthodoxy promoted by our public health officials, cutting into Pharma's bottom line in the process.

This book is dedicated to the warriors and organizations who fought the good fight for years. They went to their state houses and to Washington, DC, using their own money to fight for what was right for their children and for the safety of future generations. Many stood at the National Institutes of Health public comment microphones and begged for these studies. Many lobbied their senators and congressional representatives and sat through hearings while public health agencies spun their narratives, never committing to conduct these studies.

These mostly unintended vax-unvax studies, nested within other research, explain our government's negligence in refusing to investigate such an essential public health concern. After all, if there's nothing to hide when it comes to vaccine safety, studying health outcomes in these two populations would give credence to the health agencies' "safe and effective" mantra.

The multibillion-dollar vaccine industry and the National Vaccine Injury Compensation Program (NVICP) count on the public remaining in the dark regarding the often-devastating side effects of vaccines. Because we know the industry-captured media will censor this book, it's important to give it to friends, doctors, neighbors, expectant parents, teachers, and others. After reading *Vax-Unvax*, people will know the truth and will never be able to unknow it.

Public health officials stand in their bully pulpits promoting their versions of research and issue directives based on industry-influenced studies. Those who don't know the truth will continue to comply.

Some of the people who realized that an unprecedented cataclysm in children's health was unfolding are listed here. Many are the parents of children who became collateral damage in the war against infectious diseases. Others are courageous physicians, legislators, journalists, and researchers who realized it was time to listen to the parents and ensure that this critical research was conducted. These brave souls took a stand, vowing that no more children would suffer as the result of risky, improperly tested pharmaceutical products.

These and countless others are heroes:

Peter Aaby, MD	Sharyl Attkisson	Tom Bernard
James Adams, MD	David Ayoub, MD	Del Bigtree
Laura Fisher Andersen	Kevin Barry	Liz Birt
Lynne Arnold	Robert Scott Bell	Jennifer Black
Ed Arranga	Julia Berle	Christina Blakey
Teri Arranga	Sallie Bernard	Mark Blaxill

Kenneth Bock, MD

Charlene Bollinger

Ty Bollinger

Laura Bono

Scott Bono

Holly Bortfeld

Claire Bothwell

Judy Brasher

Sarah Bridges

Lori Brozek

Kari Bundy

Shanda Burke

Brian Burrowes

Congressman Dan Burton

Rashid Buttar, DO

Nancy Cale

Patti Carroll

Amy Carson

Stacy Cayce

Laura Cellini

Allison Chapman

Kristen Chevrier

Alan D. Clark, MD

Lujene Clark

Beth Clay

Lucy Cole

Joshua Coleman

Lou Conte

Anne Dachel

Jena Dalpez

Vicky Debold

Gayle DeLong, PhD

Richard Deth, PhD

Rosemarie Dubrowsky

Sheila Ealey, EdD

Erin Elizabeth

Norma Erikson

Becky Estepp

Barbara Loe Fisher

Wendy Fournier

Alison Fujito

David Geier

Mark Geier, MD

Patrick Gentempo, DC

John Gilmore

Eric Gladen

Dana Gorman

Doreen Granpeesheh, PhD

Becky Grant

Louise Kuo Habakus

Boyd Haley, PhD

JB Handley

Lisa Handley

Rolf Hazlehurst

Kent Heckenlively

Jackie Hines

Nancy Hokkanen

Roy Holand, MD

Mary Holland

Kristin Homme

Marcia Hooker

Shelley Hume

Suzanne Humphries, MD

Anju Iona, MD

Eileen Iorio

Jill James, PhD

Bryan Jepson, MD

Karl Kanthak

Jerry Kartzinel, MD

Janet Kern, PhD

Kelly Kerns

Rich Kerns

Heidi Kidd

David Kirby

Gary Kompothecras, DC

Robert Krakow

Arthur Krigsman, MD

Shannon Kroner, PhD

Congressman Dennis
Kucinich

Jennifer Larson

Catharine Layton

Patrick Layton

Shiloh Levine

Karri Lewis

Curt Linderman

Kim Linderman

Angela Lockhart

Tony Lyons

James Lyons-Weiler

Bobbie Manning

Leslie Manookian

Sandi Marcus

Anthony Mawson

Dr. Joe Mercola

Maureen McDonald

Karen McDonough

Lori McIlwain

Angela Medlin

Judy Mikovits, PhD

Jim Moody

Elizabeth Mumper, MD

James Neubrander, MD

James Neuenschwander, MD

Patricia Neuenschwander

Cynthia Nevison, PhD

Julie Obradovic

Dan Olmsted

Zoey O'Toole

Bernadette Pajer

Rita Palma

Katherine Paul

Leslie Phillips

Jo Pike

Sylvia Pimentel

Sunny Polito

Congressman Bill Posey

Lyn Redwood

Tommy Redwood, MD

Robert Reeves

Dawn Richardson

Bernie Rimland, PhD

Terry Roark

Wayne Rohde

Jonathan Rose

Kim Mack Rosenberg

Kim Rossi

Lenny Schaeffer

Jackie Schlegel

Beth Secosky

Barry Segal

Dolly Segal

Shelley Segal

Vera Sharav

Arnie Shreffler

Rita Shreffler

Wendy Silvers

Aaron Siri

Scott Smith, PA

Kim Spencer

Robin Rebrik Stavola

Tom Stavola Jr.

Jennifer Stella

Stephanie Stock

Kenneth Stoller, MD

Lisa Sykes

Nancy Tarlow, DC

Emily Tarsell

Ginger Taylor

Gina Tembenis

Harry Tembenis

Paul Thomas, MD

Jonathan Tommey

Polly Tommey

Toby Tommey

Yvette Negron-Torres

Bruce Vanacek

Kelly Vanacek

Brandy Vaughan

Andrew Wakefield, MBBS

Suzanne Waltman

Leslie Weed

Tim Welsh

Katie Weisman

Congressman Dave Weldon

Leah Wilcox

Theresa Wrangham

Katie Wright

Amy Yasko, PhD

Acknowledgments

Heather Ray, Margot DesBois, Sue Peters, PhD, Steven Petrosino, PhD, and Nicholas Cordeiro, NP, researched, sourced, cited, and fact-checked the manuscript with enthusiasm, tenacity, and much care. We deeply appreciate their dedication to scientific principles, accuracy, and children's health.

Zoey O'Toole, Allison Lucas, and Marcia Hooker read the manuscript and provided valuable suggestions. We are grateful for their insight, helpful recommendations, and perspective.

Our great thanks also go to Laura Bono, Jackie Hines, and Rita Shreffler for their encouragement, support, and help, especially with the final stages of this project.

We are also grateful to Tony Lyons and his team at Skyhorse Publishing, most notably Hector Carosso, for their assistance in preparing this manuscript for publication and for creating the opportunity to publish this important information.

Foreword by Del Bigtree

In May 2017, Bobby Kennedy invited Aaron Siri, Lyn Redwood, and me to a meeting with Dr. Anthony Fauci, Dr. Francis Collins, and several other public health officials at the Executive Office of the National Institutes of Health (NIH). For many years, both Bobby and I had been loudly pointing out that the Department of Health and Human Services (HHS) had undermined its duty to ensure that childhood vaccines were safe by allowing vaccine manufacturers to avoid long-term, placebo-controlled trials prior to licensure. A true vaccinated versus unvaccinated standard trial would be designed to compare a group receiving the unlicensed vaccine with a group receiving an inert, saline injection to make sure there are not any undesired health outcomes in the vaccinated group. These comparative trials are the gold standard to determine the safety of all pharmaceutical products. By the time we were meeting at NIH, sixteen vaccines had been added to the Centers for Disease Control and Prevention (CDC) recommended childhood schedule without these proper safety trials.

The CDC recommends that many of the vaccines on its schedule be given multiple times to increase effectiveness. At the time of our meeting at the NIH, most of America's children following the CDC's schedule were receiving seventy-one doses by the time they were eighteen years old. Once the CDC adds a vaccine to the recommended childhood schedule, states across the country often use their authority to mandate the vaccine for entry into school. But because the vaccines were not properly tested for safety prior to licensure, America's children were being treated as guinea pigs in a mass human experiment. Nobody knew the true risk profile of these vaccines, and nobody could say whether they were averting more problems, deaths, and illnesses than they were causing.

The best alternative to the lack of pre-licensure safety trials is to conduct post-marketing studies comparing the long-term health outcomes of vaccinated and unvaccinated individuals. Bobby and I had been outspoken about the need for these studies, which caused people like Anthony Fauci and Francis Collins to push back publicly in the mainstream media, declaring that we were deceiving and endangering the public by spreading "misinformation."

The opportunity to meet face-to-face with Fauci and Collins at the NIH was scheduled after President-elect Donald Trump asked Bobby in January 2017 to chair a new entity Trump wanted to create, the Vaccine Safety Commission. What we didn't know at the time was that Trump had accepted a million dollars from Pfizer for his inauguration. Following this, in March 2017, Trump nominated Scott Gottlieb to direct the Food and Drug Administration.[1] His nomination was subsequently approved in May 2017. Gottlieb joined Pfizer as a top executive in 2019. Additionally, Trump appointed Alex Azar as secretary of HHS after Azar had most recently served as president of the largest division of Eli Lilly. Not surprisingly, the Vaccine Safety Commission was shot down before it even got off the ground.

But there we were nonetheless, in May 2017, in a large conference room at NIH with Drs. Collins and Fauci who already had a history of calling us liars. Bobby reminded Fauci of our assertions and asked him to show us inert placebo-controlled studies for any of the seventy-one recommended vaccine doses. Fauci made a scene of going through a series of file folders that had apparently been rolled in from the NIH archives on a cart. Then, in what appeared to be feigned exasperation, he said none of the studies were there but that he would send them to us. Of course, he never did.

Aaron Siri and Bobby sent a legal demand to HHS, acting as attorneys for my group, Informed Consent Action Network (ICAN) and Bobby's Children's Health Defense (CHD), demanding it produce copies of the long-term, placebo-controlled clinical trial relied upon to license each childhood vaccine. At the same time, we also sued HHS to demand it produce copies of the biennial reports it was required to submit to Congress on how it improved the safety of childhood vaccines, and after a year of stonewalling, HHS acknowledged in a letter that this had never been done.

On June 27, 2018, HHS officially admitted in writing,

> The [Department]'s searches for records did not locate any records responsive to your request. The Department of Health and Human Services (HHS) Immediate Office of the Secretary (IOS) conducted a thorough search of its document tracking systems. The Department also conducted a comprehensive review of all relevant indexes of HHS Secretarial Correspondence records maintained at Federal Records Centers that remain in the custody of HHS. These searches did not locate records responsive to your request, or indications that records responsive to your request and in the custody of HHS are located at Federal Records Centers.[2]

The lack of HHS documents was further affirmed in a Federal Court order on July 6, 2018. We all understood how outrageous this was, but Bobby didn't rest there.

He began working with Dr. Brian Hooker to comb through all the tens of thousands of vaccine studies in the NIH official archive in

PubMed, searching for all research that compared health outcomes in vaccinated versus unvaccinated populations. And slowly, they began finding studies that either deliberately or inadvertently made these comparisons. Over the next year, Bobby and Brian published these studies one at a time on Bobby's Instagram and on CHD's website. As each study was presented, the audience was fascinated by the extraordinary and consistent results confirming that vaccinated children were unhealthier than their unvaccinated peers.

Then, in February 2021, Instagram evicted Bobby from its platform, and in August of the following year, CHD was kicked off as well. Bobby and Brian agreed they had to make the studies accessible to the public. This book is the result of their efforts.*

—Del Bigtree
CEO of Informed Consent Action Network
Host of TheHighWire.com

---

* For details on the aftermath of the May 2017 meeting with the NIH officials, see the Appendix beginning on page 193.

CHAPTER 1

Vaccinated versus Unvaccinated—Why Have the Proper Studies Not Been Conducted?

Practitioners have routinely given vaccines to children and adults since Dr. Edward Jenner developed the smallpox vaccine in 1796. In the 1940s, children received the DPT (diphtheria, pertussis, and tetanus) and smallpox vaccines; in the 1950s, children started receiving the polio vaccine; and by the late 1960s, children also received the measles, mumps, and rubella vaccines.[1] In 1986, practitioners commonly inoculated children under eighteen with eleven different shots for seven diseases. At that time, infants and children received DPT or DTaP (diphtheria, tetanus, and acellular pertussis), MMR (measles, mumps, and rubella), and polio vaccines.

Since the enactment of the 1986 National Childhood Vaccine Injury Act, which provides a liability shield for vaccine manufacturers, the vaccination schedule has multiplied considerably. Today, children following the CDC-recommended vaccination schedule receive a minimum of seventy-three shots for seventeen different diseases, with a whopping twenty-eight injections by their first birthday.[2] At a two-month "well baby visit," an infant may receive as many as six vaccines for eight different diseases.

Figure 1.1 shows a comparison of the childhood vaccination schedules in 1962, 1986, and 2023.

2023 CHILDHOOD VACCINE SCHEDULE

1962	1986	2023				
OVP	DTP (2 months)	Hep B (birth)	PCV (6 months)	Hep A (18 months)	Influenza (10 years)	
Smallpox	OVP (2 months)	Hep B (2 months)	IPV (6 months)	Influenza (24 months)	HPV (10 years)	
DTP	DTP (4 months)	Rotavirus (2 months)	COVID-19* (6 months)	Influenza (3 years)	Influenza (11 years)	
	OVP (4 months)	DTaP (2 months)	Influenza (6 months)	DTaP (4 years)	HPV (11 years)	
	DTP (6 months)	HIB (2 months)	Rotavirus (6 months)	IPV (4 years)	Tdap (12 years)	
	MMR (15 months)	PCV (2 months)	COVID-19* (7 months)	Influenza (4 years)	Influenza (12 years)	
	DTP (18 months)	IPV (2 months)	Influenza (7 months)	MMR (4 years)	Meningococcal (12 years)	
	OVP (18 months)	Rotavirus (4 months)	HIB (12 months)	Varicella (4 years)	Influenza (13 years)	
	HIB (2 years)	DTaP (4 months)	Influenza (12 months)	Influenza (5 years)	Influenza (14 years)	
	DTP (4 years)	HIB (4months)	PCV (12 months)	Influenza (6 years)	Influenza (15 years)	
	OVP (4 years)	PCV (4 months)	MMR (12 months)	Influenza (7 years)	Influenza (16 years)	
	Td (15 years)	IPV (4 months)	Varicella (12 months)	Influenza (8 years)	Meningococcal (16 years)	
		DTaP (6 months)	Hep A (12 months)	Influenza (9 years)	Influenza (17 years)	
		HIB (6 months)	DTaP (18 months)	HPV (9 years)	Influenza (18 years)	
		Hep B (6 months)				
5 Doses	**25 Doses**	**73 Doses**				

Doses are calculated based on DTaP/Tdap counting as 3 doses and MMR counting as 3 doses (as each are trivalent vaccines). The rest of the schedule is single valent. There are 6 DTaP/ Tdaps on the schedule for a total of 18 doses. There are two MMRs on the schedule for a total of 6 doses. There are 49 remaining single-valent vaccines for a total of 49+18+6 = 73 doses. *COVID-19 primary series only.

Figure 1.1—Comparison of the childhood vaccination schedules in 1962, 1986, and 2023.

Long-Term Vaccine Safety Studies Are Lacking

Despite this huge increase in vaccination, researchers have done very little to study the health of these children, either in the short term or the long term. While medical authorities credit universal childhood vaccination programs with eradicating several deadly infectious diseases, these same experts show little interest in studying the acute and long-term adverse effects of vaccination, nor do safety studies focus on the health effects of the collective vaccination schedule. Clinical trials for vaccine approval by the FDA exclusively evaluate single-vaccine products, even though infants following the CDC schedule receive up to six vaccines at the same time. Even after FDA approval, CDC completes post-market surveillance on individual vaccines only.

Many vaccines have long-term health impacts that do not become evident for years. In a 1999 interview, Anthony Fauci, former longtime director of the National Institutes of Allergy and Infectious Diseases, acknowledged that many severe injuries would remain hidden for years, and if the agency rushed vaccines to approval, "then you find out that it takes twelve years for all hell to break loose, and then what have you done?"[3]

Despite Dr. Fauci's warning, FDA clinical safety studies generally last for a relatively short duration, precluding the detection of long-term health effects. For example, researchers monitored vaccine recipients in the Engerix-B (hepatitis B) clinic trial for adverse events for only four days after injection.[4] Similarly, researchers monitored vaccine recipients in the Infanrix (DTaP) clinical trial for adverse events for only four days after injection.[5] For the ActHIB (*Haemophilus influenzae* B), scientists monitored patients for a mere forty-eight hours after injection.[6] That's it!

There is virtually no science assessing the overall health effects of the vaccination schedule or its component vaccines. In 2011, the Institute of Medicine (IOM), now the National Academy of Medicine, commissioned a committee to evaluate 158 vaccine adverse events that injury reports linked to eight different vaccines.[7] The IOM committee determined that for eighteen adverse events, evidence "convincingly supported" or "favored acceptance" of a causal relationship with administration of the vaccine.[8] The committee also determined that the relationship between five adverse events and vaccination "favored rejection."[9] However, for a colossal 135 out of the 158 adverse events/vaccine relationships considered, the IOM committee deemed the evidence "inadequate to accept or reject" the causal relationship,[10] including the relationship between the DTaP vaccine and autism. The IOM conclusion contradicts the CDC's adamant assertions that "vaccines don't cause autism."[11] Other relationships for which there is insufficient evidence of safety include the influenza vaccine and encephalopathy, the MMR vaccine and afebrile seizures, the HPV vaccine and acute disseminated encephalomyelitis, and many others. Isn't it stunning to comprehend that for almost 90% of the vaccine adverse events, CDC has never completed sufficient studies to affirm or rule out a causal relationship? This means it can't know whether these vaccines actually cause harm and certainly can't honestly say that they don't.

In 2013, the National Vaccine Program Office of the Department of Health and Human Services (DHHS) commissioned another IOM committee to update the earlier findings about the lack of evidence to support claims of safety for the entire CDC infant/child vaccination schedule.[12] The committee found that "few studies have comprehensively assessed the association between the entire immunization schedule or variations in the overall schedule and categories of health outcomes, and no study has directly examined health outcomes and stakeholder concerns in precisely the way that the committee was charged to address in its statement of task."[13] The committee continued, *"studies designed to examine the long-term effects of the cumulative number of vaccines or other aspects of the immunization schedule have not been conducted* [emphasis added]."[14] The lack of information on the overall safety of the vaccination schedule was so compelling that the committee then recommended "that the Department of Health and Human Services incorporate study of the safety of the overall childhood immunization schedule into its processes for setting priorities for research, recognizing stakeholder concerns, and establishing the priorities on the basis of epidemiological evidence, biological plausibility, and feasibility."[15] *The IOM also recommended that the CDC use its private database, the VSD, to study the overall health effects of the vaccination schedule using retrospective analyses.*[16]

A decade later, the CDC has yet to respond to the IOM committee's recommendations with a meaningful study of the health effects of the vaccination schedule.

While the CDC is not conducting these studies, what about others? Unfortunately, studying vaccine safety can come with a cost. Physicians and scientists who fall out of line with the orthodoxy of vaccinology emerge as heretics and pariahs. The most famous example took place in 1998 when Dr. Andrew Wakefield reported that 8 out of 12 of his autistic patients received the MMR vaccine prior to developing gastrointestinal symptoms and recommended further study.[17] The level of fallout was epic. Dr. Wakefield lost his medical license, reputation, and country over this brief statement he made in a now-retracted, 1998 paper in the medical journal *Lancet*. So far-reaching was his persecution that the term "Wakefielded"[18] is now used to describe the systematic gaslighting and vilification of physicians and scientists who dare to challenge vaccine orthodoxies by the government, media, and pharmaceutical enterprises. Since 1998, many other medical practitioners have paid dearly for researching vaccine risks and giving patients options that deviate from the CDC schedule. Scientists pursuing honest vaccine safety research have their peer-reviewed studies retracted and pulled out of circulation under dubious circumstances. Many have lost careers, revenue, and reputation as scientific and medical communities, government agencies, and media marginalize and condemn them.

Recently, however, the US FDA Emergency Use Authorization (EUA) for experimental, gene-based COVID-19 vaccines has illuminated for the public numerous questions about vaccine safety. Close public scrutiny of vaccine testing prompted many more people to ask tough questions. At this writing, only 69.4% of the US populace is "fully vaccinated" for COVID-19 (without accounting for boosters),[19] despite billions of dollars in advertising, systematic media propaganda, incentives, coercive measures, mandates, and numerous photo ops of government officials and celebrities receiving the shot. Officials have distributed COVID-19 vaccines in the US for approximately 30 months, and the rates of adverse events are extremely high. Medical personnel and patients have reported just over 951,000 adverse events for the vaccines (Pfizer, Moderna, Johnson & Johnson, and Novavax) in the US alone.[20] In fact, in three years, COVID-19 shots have caused 97% of *all* adverse events reported to the CDC's Vaccine Adverse Events Reporting System (VAERS) since the introduction of this program in 1986. The media are now beginning to acknowledge certain adverse events, albeit with the obligatory disclaimer regarding how "rare" vaccine injuries are.

Why Aren't the Necessary Studies Being Conducted?

One reason regulators give to dismiss a more rigorous approach in studying the long-term health effects of the vaccination schedule is that vaccine adverse events are "one-in-a-million," and thus we should stop promoting fear of vaccine injury. The government derives its one-in-a-million figure by comparing the number of compensated vaccine injuries by the National Vaccine Injury Compensation Program (NVICP) to the total number of vaccines given in the US.[21] Unfortunately, most vaccine-injured people don't even know the NVICP exists, and even fewer get compensated.[22] The Lazarus study, which CDC funded and then abandoned—likely because the agency didn't like the results—stands in stark contrast to the one-in-a-million figure. Specifically, researchers in the Lazarus study found the rate of adverse events to be 1 in 38[23] among a population of about 375,000 individuals given 1.4 million routine vaccines. Over the three-year study period, that translated to an individual having a 1 in 10 chance of experiencing an adverse reaction to a vaccine. This is a far cry from the mythical "one-in-a-million" rhetoric touted by the pharmaceutical industry and government health agencies. The Lazarus study suggests that federal officials and the pharmaceutical industry must pay urgent attention to this astronomical rate of adverse events. Still, the CDC and FDA steadfastly refuse to study health outcomes in vaccinated versus unvaccinated populations.

Viable Options for a Vax versus Unvax Study Are Available

A randomized controlled trial (RCT) is a prospective study (looking at health effects in the future) wherein researchers randomly choose individuals from a pool of volunteers to make up either the experimental or control group. Then, researchers blind both groups to what they've received (treatment or placebo) to avoid bias among the trial participants.

In FDA clinical trials, the experimental group receives the vaccine, and the control group receives the placebo. CDC guidance requires a placebo to be physiologically inert, like a saline solution. However, most vaccine clinical trials lack a true saline placebo, making a proper evaluation of vaccine safety impossible. For example, the FDA did not require an inert placebo prior to its 2007 approval of the Gardasil® human papillomavirus vaccine. In fact, rather than using a saline placebo, researchers gave the control group an injection of highly toxic amorphous aluminum hydroxyphosphate sulfate (AAHS), a strong adjuvant[24] with no prior safety testing.[25] Then in the follow-up trial for Merck's Gardasil-9 vaccine, approved in 2014, researchers gave the original Gardasil® vaccine as the placebo control.[26] As another example, in a study of flu vaccines in pregnant women, researchers gave the control group a meningococcal vaccine that the FDA has never tested for safety in pregnancy.[27]

Public health experts assert they can't feasibly study vaccinated versus unvaccinated populations because it would be unethical to complete a RCT where researchers withhold lifesaving vaccines from a blinded placebo control group.[28] Their argument is a sham. Pharmaceutical companies typically use this method during the FDA approval process to test new drugs or biologics when no comparable treatment exists. For example, FDA requires RCT clinical studies for certain cancer treatments,[29,30] heart medications,[31] and respiratory drugs,[32] and no one appears to question the ethics of withholding potentially lifesaving remedies from blinded placebo control groups. It is, in fact, standard practice.

Yet when a medical journalist conducting a March 23, 2015, interview with Frontline asked Dr. Paul Offit, the Director of the Vaccine Education Center at Children's Hospital of Philadelphia and a vocal defender of the vaccine industry, about an RCT between vaccinated and unvaccinated children to determine whether vaccines cause autism, the doctor stated, "It is highly unethical to do a study like that."[33] He explained that such a study would have "frankly condemned those in the unvaccinated group—some in the unvaccinated group— to develop diseases which can permanently harm them and/or kill them."[34] Additionally, the Children's Hospital of Philadelphia's "Ethical Issues and Vaccines" website states, regarding vaccine safety testing, that "failing to provide any adequate prevention option (to the control group) can be a difficult decision when the vaccine can potentially prevent a serious, untreatable, or fatal infection."[35]

The fact that vaccine proponents apply this flawed rationale to vaccines alone and not other medicine suggests an agenda not rooted in science or logic. Furthermore, researchers can complete many other types of analyses besides RCT using existing populations of vaccinated and unvaccinated children and adults that, according to the Cochrane Collaboration,[36] produce results equal in reliability.[37] These include analyses that are prospective (looking at health effects in the future) or retrospective (looking at past medical data and history). In fact, CDC scientists routinely complete unblinded, retrospective vaccine safety studies (i.e., not RCTs). Furthermore, the CDC often touts these types of studies regarding the MMR vaccine[38] and thimerosal-containing vaccines[39,40] as proof that vaccines do not cause autism. These studies are all based on retrospectively compiled datasets, including the CDC's own Vaccine Safety Datalink (VSD).[41] The VSD is a compilation of data from nine health maintenance organizations (HMOs), including over two million children. The CDC's VSD also contains records for unvaccinated children, making it an ideal data source for assessing vaccine safety. And yet CDC scientists have never performed a retrospective vax-unvax comparison study.

Another excuse for vaccination versus unvaccinated studies not being conducted is that the medical establishment tells us that groups of unvaccinated children are so unique that researchers couldn't legitimately compare them to vaccinated children in scientific studies. For example, in response to UPI reporter Dan Olmsted reporting on the nonexistence of autism in Amish children (who are unvaccinated), Dr. Offit stated, "you're selecting for two very different groups of people when you choose children who are completely vaccinated or completely unvaccinated. It would be hard to control for that."[42] The medical establishment claimed—without evidence—that the Amish are a unique, genetically distinct population that shouldn't be compared to other groups.[43] This argument is flawed because while the Amish may or may not be genetically different, they make up only a small portion of the unvaccinated in the US. For example, in a survey completed by the CDC in 2015, 1.3% of all 24-month-olds had yet to receive a single vaccine from the CDC's infant schedule.[44] Yet the Amish account for only about 0.08% of the US population.[45] Therefore, even if researchers excluded the Amish from study, there are plenty of unvaccinated children and adults for this type of research, beyond small pockets of potentially "genetically distinct" populations.

Purpose of This Book

Before the pandemic, we began searching for publications in which researchers studied the health outcomes of vaccinated versus unvaccinated populations. We have so far identified over 100 peer-reviewed articles from open, peer-reviewed, scientific, and medical literature. In addition, many other research papers support the conclusions of these studies. This book is a compendium of these studies.[46] We also included relevant research studies published by other reputable sources.

We wrote this book for parents, curious laypersons, and anyone concerned with protecting children's health. In the following chapters, we summarized each of the "vax versus unvax" studies, included bar graphs that illustrate the most pertinent results, and organized chapters around different vaccines and vaccine components. By simply paging through the chapters, you will understand the different outcomes associated with the vaccination schedule and individual vaccines within it. We also hope you develop an appreciation of the complexities of vaccine safety science beyond the very simplistic picture that health officials and the media customarily paint.

Statistical Terminology Explained

To assist you, we offer a brief primer in epidemiology since most of the studies this book reviews are epidemiological. Terms including "odds ratio," "relative risk," and "hazard ratio" are key concepts for understanding these studies. These terms are all different ways of expressing the likelihood of having a disorder in the vaccinated group versus the likelihood of having the same disorder in the unvaccinated group.

- **Odds ratio** is a way of expressing these likelihoods or "odds" based on the proportion of individuals in each group who have the disorder versus those who don't. For example, an odds ratio of 2.0 for developmental delays in vaccinated versus unvaccinated means that the proportion of individuals who possess developmental delays is twice as high in the vaccinated group compared to the unvaccinated group.
- **Relative risk** is a ratio of the risk of the disorder in the vaccinated group versus that in the unvaccinated group. For example, a relative risk of 2.0 for developmental delays means that the proportion of people with developmental delays versus the whole sample of vaccinated people (both those who do and don't have developmental delays) is twice as high in the vaccinated group.
- **Hazard ratios** are used less frequently in epidemiology and represent more of a measure of "instantaneous risk," whereas when researchers calculate odds ratios and relative risk, the "odds" or "risk" is calculated cumulatively over the entire duration of the study. For example, exactly five years after vaccination, the hazard ratio of experiencing a particular adverse event might be 2.0 compared to the unvaccinated. However, the cumulative risk averaged over that period (i.e., from vaccination to five years after vaccination) might differ, say, 3.0. The former value is a hazard ratio, and the latter is a relative risk.
- **P-value** or probability value measures the likelihood that a particular relationship is produced by random chance rather

than a true correlation, on a scale of 0 to 1. A p-value of 1.0 would imply a completely random result supporting the "null hypothesis." The null hypothesis means no relationship between "x" and "y" exists. A p-value approaching zero shows a strong relationship between "x" and "y" (e.g., "vaccination" and "adverse event"). The gold standard for achieving statistical significance is when the p-value is less than 0.05, meaning less than a 5% chance that the correlation was random. Of course, p-values much lower than 0.05 give additional confidence in a strong correlation, as the calculated p-value can be as low as <0.0001.

- **95% confidence interval** or **95% CI** is an alternative to p-value. This consists of two numbers that bracket the actual odds ratio, relative risk, or hazard ratio. For example, let's say the relative risk of asthma is 1.5 in the vaccinated group versus the unvaccinated group with a 95% CI of 1.1 to 1.9. This would mean that we are 95% confident that the true relative risk in the analysis is somewhere between the bounds 1.1 and 1.9. Also, because the lower bound is 1.1 and does not cross a value of 1.0, we would consider this result to be statistically significant (like a p-value of less than 0.05). In other words, we are 95% confident that the relative risk is at least 1.1. Once the lower bound dips below 1.0, statistical significance is not achieved because 1.0 means there is no difference between the outcome between the vaccinated and the unvaccinated. Like lower p-values (i.e., much lower than 0.05), 95% CIs that tightly bracket the calculated value of odds ratio or relative risk *and* that are well above the lower bound of 1.0 give additional confidence that a relationship is significant and not achieved by random chance.

CHAPTER 2

Health Outcomes Associated with the Vaccination Schedule

Despite the call of the 2013 IOM Committee to investigate the health effects of the childhood vaccination schedule,[1] researchers have conducted few studies regarding health outcomes associated with the entire schedule. In fact, FDA and CDC scientists have not completed a single analysis. Instead, private grants and foundations funded this research. This chapter highlights studies primarily found in peer-reviewed scientific literature that looked at health outcomes associated with the vaccine schedule. We also present supporting research published elsewhere. University professors from Vanderbilt University, Jackson State University, and the University of Chicago, as well as medical practitioners, independent scientists, and analysts, authored these studies.

Figure 2.1 shows the results from the paper "Pilot Comparative Study on the Health of Vaccinated and Unvaccinated 6- to 12-Year-Old U.S. Children," published in the *Journal of Translational Sciences* in 2017[2] (First Mawson Study). The paper's lead author, Dr. Anthony Mawson, is a professor in the Department of Epidemiology and

Pilot Comparative Study on the Health of Vaccinated
and Unvaccinated 6- to 12-Year-Old U.S. Children

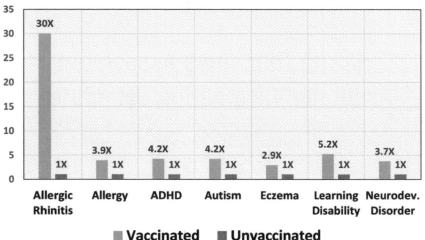

Figure 2.1—The odds ratios of chronic diseases for vaccinated
versus unvaccinated children (Mawson et al. 2017a).

Biostatistics at the School of Public Health at Jackson State University
in Jackson, Mississippi. The First Mawson Study was the first peer-reviewed, published study to consider the health effects of the entire
vaccination schedule on children. The authors surveyed the parents of
666 homeschooled children, including 261 completely unvaccinated
children. Eighty-eight percent of the children in the study were white,
with an average age of nine years, and 52% of the participants were
female.

The authors found that vaccinated children, including fully and
partially vaccinated groups, had significantly fewer chicken pox
and pertussis cases.[3] However, as indicated in Figure 2.1, vaccinated
children had 30 times greater odds of a diagnosis of allergic rhinitis (p-value <0.001, 95% CI of 4.1 to 219.3), 3.9 times greater odds of

allergies (p-value <0.001 and a 95% CI of 2.3 to 6.6), 4.2 times greater odds of attention-deficit/hyperactivity disorder (ADHD) (p-value = 0.013 and a 95% CI of 1.2 to 14.5), 4.2 times greater odds of autism (p-value = 0.013 and a 95% CI of 1.2 to 14.5), 2.9 times greater odds of eczema (p-value = 0.035 and a 95% CI of 1.4 to 6.1), 3.7 times greater odds of neurodevelopmental disorders (p-value < 0.001 and a 95% CI of 1.7 to 7.9), and 5.2 times greater odds of learning disabilities (p-value = 0.003 and a 95% CI of 1.6 to 17.4).[4] These odds ratios are all statistically significant. Compared to 197 children in the fully vaccinated and 261 in the unvaccinated groups, the 208 children in the partially vaccinated group achieved "an intermediate position regarding allergic rhinitis, ADHD, eczema, and learning disability."[5]

Figure 2.2 shows the percentage of children diagnosed with pneumonia and ear infections in the vaccinated and unvaccinated groups in

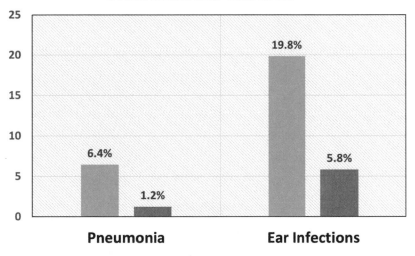

Figure 2.2—Rates of infections reported in vaccinated versus unvaccinated children (Mawson et al. 2017a).

the First Mawson Study. The researchers found that 6.4% of vaccinated children had been diagnosed with pneumonia versus 1.2% of unvaccinated children (p-value < 0.001 and a 95% CI of 1.8 to 19.7).[6] Likewise, 19.8% of vaccinated children were diagnosed with ear infections versus just 5.8% of unvaccinated children (p-value < 0.001 and a 95% CI of 2.1 to 6.6).[7] The differences between the two groups were statistically significant, as p-values were less than 0.005.

Figure 2.3 shows results from the paper "Preterm Birth, Vaccination and Neurodevelopmental Disorders: A Cross-Sectional Study of 6- to 12-Year-Old Vaccinated and Unvaccinated Children," published in the

Preterm Birth, Vaccination and Neurodevelopmental Disorders: A Cross-Sectional Study of 6- to 12-Year-Old Vaccinated and Unvaccinated Children

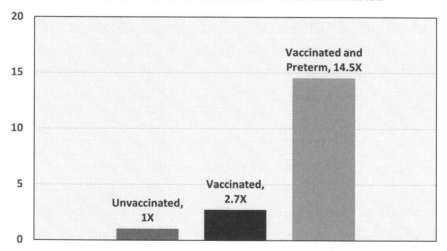

Figure 2.3—Odds ratios of a neurodevelopmental disorder diagnosis for unvaccinated, vaccinated, premature, and premature birth/vaccinated children (Mawson et al. 2017b).

Journal of Translational Science in 2017[8] (Second Mawson Study). Dr. Anthony Mawson was also the lead author of this study. Using the First Mawson Study dataset, researchers performed a follow-up study, employing a different statistical model to adjust for significant factors, including gender, adverse environments, medications, and vaccines during pregnancy. As a result, the Second Mawson Study found that vaccinated children were diagnosed with 2.7 times greater odds of a neurodevelopmental disability (NDD) than their unvaccinated peers (p-value = 0.012 and a 95% CI of 1.2 to 6.0).[9] Also, vaccinated preterm children were diagnosed with NDDs with 14.5 times greater odds (p-value < 0.001 and a 95% CI of 5.4 to 38.7) when compared to a reference group of unvaccinated children born at full gestational term.[10]

The two Mawson Studies are the first of their kind, published four years after the IOM's call for additional research on the childhood vaccination schedule.[11] The journal, *Frontiers in Public Health*, accepted the First Mawson Study in early 2017. *Frontiers in Public Health* is a highly reputable journal indexed in PubMed, a search engine with access to over 34 million citations for biomedical literature.[12] PubMed was developed in 1996 and is maintained by the National Center for Biotechnology Information and the US National Library of Medicine, supported by the National Institutes of Health.[13] It is the gold standard for medical literature.

The journal posted the First Mawson Study abstract on its website, and it received immediate attention on social media. It was viewed over 80,000 times the first weekend after posting. However, due to backlash regarding the article's subject matter, the journal unceremoniously removed the abstract after just three days and rescinded its acceptance of the paper, dealing a harsh blow to Dr. Mawson and his coauthors. Still, the journal's action did not constitute a full retraction, as editors stated that the journal had initially only provisionally accepted the paper. Conversely, a retraction of an article removes an already published paper from a journal. Retractions may occur because of errors in the research, issues of reproducibility, plagiarism, falsification of data or results, fabrication of data or results, copyright infringement, or failure to disclose conflicts of interest.[14] Unfortunately, forced retractions have become a tool to smear studies with none of the above defects simply because they contain unfavorable or unpopular results.

Dr. Mawson received an email from the journal's chief editor, Dr. Joav Merrick, stating that *Frontiers* could not accept the paper for publication based on several issues inherent to survey-based studies. First, Dr. Merrick argued that the survey response rate was not verifiable. This was true, as the survey was available online nationwide for three months, leaving no way to ascertain a response rate. However, the journal did not raise this issue during the rigorous, original peer review, nor did it warrant retraction or withdrawal of the paper based on the Committee on Publication Ethics guidelines.[15] The chief editor also complained that authors could not verify medical diagnoses, but again, this is inherent to existing published, peer-reviewed, survey-based studies. If the journal had indeed found this unacceptable, it would have raised the issue in the initial peer review of the paper.

Subsequently, the *Journal of Translational Science*, a highly reputable, peer-reviewed scientific journal (though not indexed in PubMed), published the two Mawson Studies.[16, 17] The authors of the Mawson Studies went this route, as it was the only option to get the results in print, in a peer-reviewed, science-based journal. Unfortunately, there is still a substantial lack of similar literature, especially in PubMed journals, and more work is desperately needed in this area.

Figure 2.4 shows the results from the paper "Analysis of Health Outcomes in Vaccinated and Unvaccinated Children: Developmental Delays, Asthma, Ear Infections and Gastrointestinal Disorders," published in the journal *SAGE Open Medicine* in 2020.[18] The lead author, Dr. Brian Hooker, is professor emeritus of biology at Simpson

Analysis of Health Outcomes in Vaccinated and Unvaccinated
Children: Developmental Delays, Asthma, Ear Infections
and Gastrointestinal Disorders

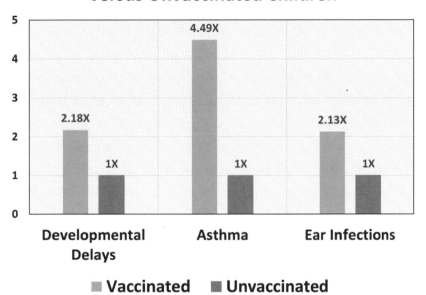

Figure 2.4—Odds ratios of diagnosed disorders in vaccinated versus
unvaccinated children: Developmental delays, asthma, and ear infections
(Hooker and Miller, 2020).

University in Redding, California. In this study, Hooker and his coauthor, Neil Miller, examined the medical records from three different pediatric practices in various US areas. The authors followed 2,047 patients from birth to a minimum age of three years and a maximum age of 12.5 years. They divided the children into two groups: those who received any vaccines before their first birthday (69.1%) and those who did not (30.9%). The authors accounted for diagnoses only after a child's first birthday to establish that vaccination preceded the first diagnosis of illness or disorder. As shown in Figure 2.4, vaccinated children were diagnosed with developmental delays at 2.18 times greater odds (p-value < 0.0001 and a 95% CI of 1.47 to 3.24), asthma at 4.49 times greater odds (p-value = 0.0002 and a 95% CI of 2.04 to 9.88), and ear infections at 2.13 times greater odds (p-value < 0.0001 and a 95% CI of 1.63 to 2.78) than unvaccinated children.[19] These odds ratios were statistically significant.

The authors also assessed a control diagnosis of head injury to establish whether differences in diagnoses between the two groups were due to disparities in healthcare-seeking behavior between vaccinating and non-vaccinating families.[20] In other words, do vaccinated children tend to visit their doctor more than unvaccinated in this cohort? Children within vaccinated and unvaccinated groups should not have different head injury incidences unrelated to vaccination status. Otherwise, the authors would need to control this in the statistical model. However, vaccinated and unvaccinated groups did not have a statistically different incidence of head injury, which affirmed the validity of the other results.

In a separate analysis, Hooker and Miller changed the age range of children in the cohort to between five years and 12.5 years of age. By raising the minimum age from three to five, these researchers allowed diagnoses not typically made at a younger age to emerge. As shown in Figure 2.5, within this age group, vaccinated children had 2.48 times greater odds (p-value = 0.045 and a 95% CI of 1.02 to 6.02) of a gastrointestinal disorder than unvaccinated children.[21] This result was statistically significant. Vaccinated children also had significantly greater odds of asthma, ear infections, and developmental delays than unvaccinated children.[22]

According to the study authors, five medical journals rejected the paper outright, without peer review, before *SAGE Open Medicine* considered it. *SAGE* took eleven months to complete the peer review as the journal editors had to search for scientific peers willing to evaluate the manuscript. Unfortunately, many declined the invitation. Once

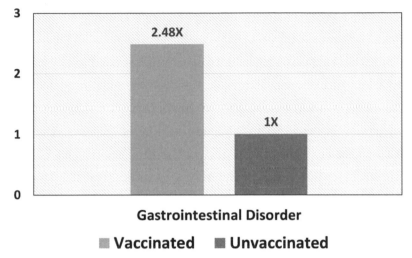

Odds Ratios of Gastrointestinal Disorder in Vaccinated versus Unvaccinated Children

Gastrointestinal Disorder

■ **Vaccinated**　■ **Unvaccinated**

Figure 2.5—Odds ratios of a diagnosis of gastrointestinal disorder in vaccinated versus unvaccinated children (Hooker and Miller, 2020).

the journal identified reviewers, these scientific peers reviewed the paper three times with revisions before accepting the manuscript. This arduous process is not typical, as most journals require only one round of peer review. Still, the result was a solid stake in the ground regarding the health outcomes of vaccinated versus unvaccinated children in a reputable medical journal, *SAGE Open Medicine*, which is indexed in PubMed.

The paper has been viewed or downloaded over 200,000 times, and the journal has not retracted it. Vaccine zealots who might take issue with the findings, for the most part, have refrained from attacking the study. The paper did fall victim to "fact-checking," however. The organization Health Feedback, which works with Facebook, claimed that the findings were "unsupported," and this assertion appears whenever someone posts a link to the paper on the social media platform.[23] The "fact-checkers" took issue with the convenience sample nature of the study and alleged that the three medical practices included in the study were not a representative sample of the US population. When challenged with a rebuttal by the authors, who presented multiple other reputable studies based on convenience samples, the "fact-checkers" were silent and ultimately chose to ignore the sound logic presented by the study authors. The level of censorship was so blatant, and without recourse, Children's Heath Defense filed a civil lawsuit against Facebook for this and other instances of capricious editing or removal of author posts.

Figure 2.6 shows results from the paper "Health Effects for Vaccination versus Unvaccinated Children, with Covariates for Breastfeeding Status and Type of Birth," published in the *Journal of Translational Science* in 2021.[24] In this follow-up paper (2021 Study), Hooker and Miller investigated three additional pediatric practices that conducted parent surveys of children's vaccination status, demographic factors, and medical diagnoses. With access to medical records, the study authors confirmed children's survey results by reviewing charts for individual patients participating in the study. There were 1,565 children in the total sample: 60.4% were unvaccinated, 30.9% were partially vaccinated, and 8.7% were up to date on their vaccinations.[25]

The authors also considered other factors in the analysis, including whether participants were breastfed for at least six months, born vaginally or by cesarean section, and homeschooled or attended a public or private school. Hooker and Miller found significantly higher levels of severe allergies, autism, asthma, gastrointestinal disorders, ADD/ADHD, and chronic ear infections among fully vaccinated children compared to their unvaccinated peers.[26] As compared to the 2020 study,[27] the 2021 study showed that fully vaccinated children had much higher odds ratios of asthma (17.6 versus 4.49, with a p-value

Health Effects for Vaccination versus Unvaccinated Children, with Covariates for Breastfeeding Status and Type of Birth

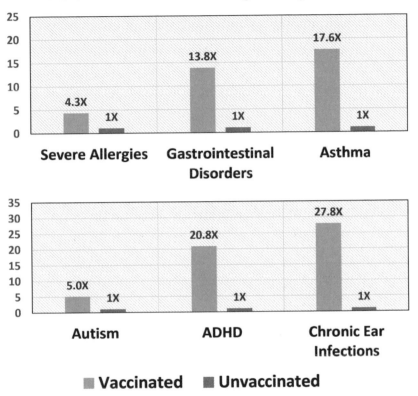

Odds Ratio for Obtaining a Diagnosis

Figure 2.6—Odds ratios for severe allergies, gastrointestinal disorders, asthma, autism, ADHD, and chronic ear infections in vaccinated versus unvaccinated children (Hooker and Miller, 2021).

<0.0001 and a 95% CI of 6.94 to 44.4), gastrointestinal disorders (13.8 versus 2.48, with a p-value <0.0001 and a 95% CI of 5.85 to 32.5), and ear infections (27.8 versus 2.13, with a p-value < 0.001 and a 95% CI of 9.56 to 80.8).[28] In the 2021 study, the authors compared fully vaccinated children to fully unvaccinated children. The 2020 study compared fully and partially vaccinated children to unvaccinated children.[29, 30] Finally, in the 2021 study, vaccinated children were diagnosed with

chicken pox significantly less frequently than unvaccinated children.[31] This expected result helped confirm the legitimacy of the 2021 study analysis.

Figure 2.7 shows results from the 2021 study by Hooker and Miller. In looking at the combined effect of vaccination and breast-feeding, Hooker and Miller found that unvaccinated children who were breastfed for at least six months were diagnosed significantly less frequently with severe allergies, autism, asthma, gastrointestinal disorders, ADD/ADHD, and chronic ear infections compared to vaccinated, non-breastfed peers.[32] Figure 2.7 shows an example (for asthma) of the increasing odds ratios observed for each condition studied. With unvaccinated/breastfed babies as the reference group, unvaccinated/

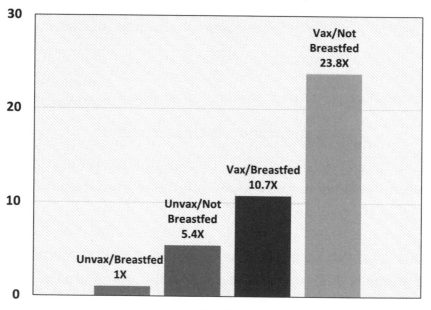

Figure 2.7—Odds ratios for a diagnosis of asthma within vaccinated and unvaccinated children accounting for breastfeeding status (Hooker and Miller, 2021).

non-breastfed babies had 5.4 times greater odds of an asthma diagnosis (p-value = 0.040), vaccinated/breastfed babies had 10.7 times greater odds (p-value < 0.0001), and vaccinated/non-breastfed babies fared the worst with 23.8 times greater odds (p-value < 0.0001) of an asthma diagnosis.[33] Unvaccinated children born vaginally were also diagnosed less frequently with severe allergies, autism, asthma, gastrointestinal disorders, ADD/ADHD, and chronic ear infections than vaccinated children born via cesarean section (results not shown).[34] Homeschooled children showed no difference in the outcomes measured compared to those who attended public or private schools.[35]

Figure 2.8 shows results from the paper "Relative Incidence of Office Visits and Cumulative Rates of Billed Diagnoses along the Axis of Vaccination," appearing in the *International Journal of Environmental Research and Public Health* in 2021.[36] The lead author, James Lyons-Weiler, PhD, is the director of the Institute for Pure and Applied Knowledge in Pittsburgh, Pennsylvania. Lyons-Weiler and his coauthor, Dr. Paul Thomas, took a unique approach to investigating health differences between vaccinated and unvaccinated children in Dr. Thomas's Portland, Oregon, medical practice. Rather than examining whether children had ever been diagnosed with the disorders studied, they compared the number of office visits associated with specific diagnoses for vaccinated versus unvaccinated children. This comparison, termed "Relative Incidence of Office Visits" (RIOV), reflected the number of times the physician saw children diagnosed with the disorder for vaccinated children versus unvaccinated children.[37] Lyons-Weiler wrote, "Our measure, the Relative Incidence of Office Visits (RIOV), is sensitive to the severity of disease and disorder—specifically, the disease burden."[38] RIOV also reflects the frequency of recurring diseases such as fever, ear infections, and respiratory infections.

In this assessment of 2,763 fully and partially vaccinated children and 561 unvaccinated children, as shown in Figure 2.8, vaccinated

Relative Incidence of Office Visits and Cumulative Rates of Billed Diagnoses along the Axis of Vaccination

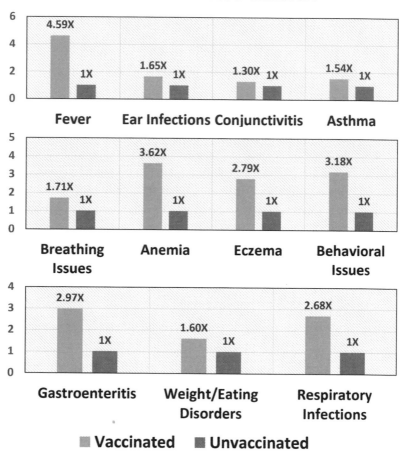

Figure 2.8—Ratio of pediatrician office visits for listed disorders between vaccinated and unvaccinated children (Lyons-Weiler and Thomas, 2021).

children had significantly more office visits associated with ear infections, conjunctivitis, breathing issues, anemia, eczema, behavioral issues, gastroenteritis, weight and eating disorders, and respiratory infections than unvaccinated children.[39] Unvaccinated children experienced more chicken pox and pertussis.[40] Due to meager rates of certain conditions in Dr. Thomas's practice, such as developmental disorders, the researchers could not determine statistically significant differences between the two groups for those conditions. However, the authors remarked that zero unvaccinated patients exhibited ADHD compared to 5.3% of the vaccinated group.[41]

Dr. Thomas opened his Portland, Oregon, pediatric practice in 2008 to improve children's health by providing individualized, holistic medical care based on informed consent.[42] Unlike most pediatric practices in the US, Dr. Thomas offered parents the option to delay or modify their child's vaccination schedule. These alternative schedules reduced a child's exposure to toxic vaccine components and allowed skipping one or more vaccines in response to relevant factors such as the history of autoimmunity, vaccine injury, or other personal choices. Dr. Thomas's practice rapidly grew to over 15,000 patients, with a staff of more than thirty. In 2016, Dr. Thomas published his clinical approach to vaccination with coauthor Jennifer Margulis, PhD, in the bestselling book *The Vaccine-Friendly Plan: Dr. Paul's Safe and Effective Approach to Immunity and Health—From Pregnancy through Your Child's Teen Years.*[43] Despite the Oregon Medical Board's repeated letters of complaint, threats, and termination of health insurance contracts related to his refusal to strictly follow the CDC vaccination schedule, Dr. Thomas honored and served his patients according to the Hippocratic oath and his scientifically informed and experience-based understanding of children's health.[44]

In February 2019, Dr. Thomas received a letter from the Oregon Medical Board requesting that he scientifically demonstrate that The Vaccine-Friendly Plan, an alternative vaccination schedule laid out in his book, was as safe as the CDC's immunization schedule for children, despite the CDC's inability to produce valid scientific evidence of the safety of their own schedule.[45] Thomas recognized that his practice's mixture of entirely, partially, and unvaccinated children comprised a uniquely valuable pool of clinical data to investigate and compare health outcomes based on vaccination.[46] Therefore, he hired an independent analyst to perform a quality assurance analysis on his practice and worked with research scientist James Lyons-Weiler, PhD, to analyze the data and write a report.[47] Regarding his decision to conduct this vaccinated versus unvaccinated study, Dr. Thomas said, "Because no other practice in town is doing this, I'm uniquely sitting on a population and get to see the difference. And that's why I felt it was my ethical obligation to publish this data so the world would know."[48]

Rather than acknowledge Dr. Thomas's thorough response to its challenge, the Oregon Medical Board issued an "emergency order" to suspend Dr. Thomas's medical license within a week of the publication of this research.[49] The order stated that Dr. Thomas's "continued practice constitutes an immediate danger to the public" and his breach of the "standard of care has placed the health and safety of many of his patients at serious risk of harm."[50] The Board's letter declared that Thomas touts his alternative vaccination plan as "providing superior results to any other option, namely improved health on many measures" and "fraudulently asserts that following his vaccine schedule will prevent or decrease the incidence of autism and other developmental disorders."[51] It accused him of using "this claim to solicit parental 'refusal' of full vaccination for their children, thereby exposing them to multiple potentially debilitating and life-threatening illnesses, including tetanus, hepatitis, pertussis (whooping cough), rotavirus, measles, mumps, and rubella."[52] The order outlined several largely contrived cases to falsely accuse Dr. Thomas of "unprofessional or dishonorable conduct" and medical negligence for declining to vaccinate children or advise parents in strict accordance with the CDC guidelines.[53] Journalist Jeremy Hammond detailed Dr. Thomas's story in the book *The War on Informed Consent: The Persecution of Dr. Paul Thomas by the Oregon Medical Board.*[54]

The Oregon Medical Board reinstated Dr. Thomas's license in June 2021 under the conditions that Thomas only treat patients needing acute care, that he refrain from consulting with parents or patients, or directing or instructing clinic staff "relating to vaccination protocols, questions, issues, or recommendations," and that he conduct no further research related to patient care.[55] Because of these limitations, the Board effectively prevented Dr. Thomas from providing his high standard of care and educating families about their medical options. While disinclined to accept these terms, Dr. Thomas agreed to them in order to keep his practice afloat amidst the financial sanctions the Board had already imposed.[56] Subsequently, rather than suffer through a long, expensive fight to overturn the Board's conditions, Dr. Thomas relinquished his medical license on December 6, 2022, and retired from practice.[57]

Under dubious circumstances, The *International Journal of Environmental Research and Public Health* editors retracted the Lyons-Weiler and Thomas paper in July 2021.[58] The retraction statement included a brief, vague explanation: "Following publication, concerns were brought to the attention of the editorial office regarding the validity of the conclusions of the published research. Adhering to our complaints procedure, an investigation was conducted that raised several methodological issues and confirmed that the conclusions were not supported by strong scientific data."[59] According to lead author Lyons-Weiler, the journal's decision to retract the paper was based on a lone, anonymous complaint regarding an alternative explanation of the statistical results. The journal publicly offered no details of the methodological issues cited, and the article remains unpublished. According to Lyons-Weiler, the complaint alleged that any differences between vaccinated and unvaccinated children were due to differences in healthcare-seeking behavior, i.e., that vaccinated individuals see their medical practitioners more frequently. However, this allegation has been thoroughly refuted specifically for the case of Dr. Thomas's practice in the follow-up publication, "Revisiting Excess Diagnoses of Illnesses and Conditions in Children Whose Parents Provided Informed Permission to Vaccinate Them," by Lyons-Weiler and Dr. Russell Blaylock, published in the *International Journal of Vaccine Theory, Practice, and Research* in 2022.[60]

Figure 2.9 shows the results from a self-published study completed in 2004 by the Dutch Association for Conscientious Vaccination in

Dutch Association for Conscientious Vaccination Self-Published Study

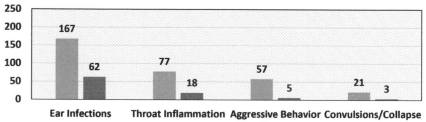

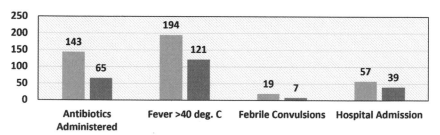

■ Vaccinated   ■ Unvaccinated

Figure 2.9—Number of diagnoses per 100 children in the first five years of life for vaccinated versus unvaccinated children (Diseases and Vaccines: NVKP Survey Results).

Driebergen, the Netherlands.[61] The study authors investigated the respective health outcomes of 312 vaccinated and 231 unvaccinated children. All vaccinated children followed the recommended schedule of the Dutch Vaccination Program, and the study authors excluded partially vaccinated children from the study. The Dutch infant vaccination schedule includes only six vaccines given prior to the first birthday.[62] However, three of the vaccines are hexavalent, meaning one injection contains antigens for six different diseases. Figure 2.9 compares the incidence of acute conditions per 100 children, and Figure 2.10 below compares the incidence of chronic conditions per 100 children.

Although vaccination protected against vaccine-preventable infectious diseases such as pertussis and measles, vaccinated children had a far greater incidence of behavioral issues, seizures, loss of consciousness, antibiotic use, and hospital visits, among others.[63] Results for ear infections and fever are consistent with the results discussed previously in this chapter.[64, 65, 66, 67, 68] Eight of the vaccinated participants and none of the unvaccinated children were diagnosed with autism.[69] This outcome is consistent with the results obtained in the Mawson Studies,[70] by Hooker and Miller,[71] and Lyons-Weiler and Thomas,[72] who also observed lower autism incidence in unvaccinated children.

Figure 2.10 shows results from the previous Dutch study that are consistent with those previously discussed for asthma,[73, 74, 75] allergies,[76, 77] and eczema.[78, 79]

Joy Garner founded The Control Group and authored the report "The Control Group Pilot Survey of Unvaccinated Americans," released on February 9, 2021.[80] Joy Garner is a tech inventor for video game hardware and a US patent holder. The survey included 1,482 participants, with 1,272 children, from 48 US states. Parents provided survey data for children. The Control Group's statisticians compiled

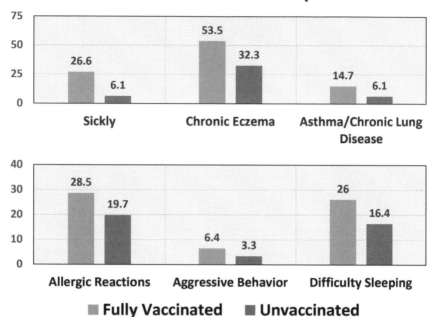

Figure 2.10—Incidence of diagnosis per 100 children for vaccinated versus unvaccinated children (Diseases and Vaccines: NVKP Survey Results).

disease incidence data for completely unvaccinated survey participants and compared the information to disease incidences for the entire US population (obtained through federal sources, including the CDC and NIH). The Control Group assumed that US disease incidence data reflected the vaccinated population of the US since, according to their study, 99.74% of Americans are vaccinated.[81]

Figure 2.11 shows substantially higher levels of single and multiple chronic disorders among vaccinated versus unvaccinated American children. Data for vaccinated children are based on the CDC report *Preventing Chronic Disease* and do not include diagnoses of obesity.[82]

The Control Group Pilot Survey of Unvaccinated America

Single and Multiple Chronic Conditions Among Vaccinated and Unvaccinated U.S. Children

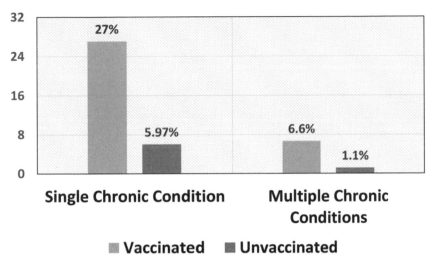

Figure 2.11—Comparison of percentage of US children with a single chronic condition or multiple chronic conditions among vaccinated versus unvaccinated (Statistical Evaluation of Health Outcomes of the Unvaccinated, Joy Garner, The Control Group, February 9, 2021, thecontrolgroup.org).

Figure 2.12 shows results also taken from The Control Group study and demonstrates that vaccinated children have a much higher incidence of specific chronic disorders than unvaccinated children.[91] Most notably, the vaccinated have a twenty times higher incidence of ADHD than the unvaccinated (9.4% versus 0.47%) and over a ten times higher incidence of autism than the unvaccinated (2.5% versus 0.21%).[92] This is consistent with the study by Hooker and Miller, which found odds ratios for ADD/ADHD and autism of 20.8 and 5.0 (p-value < 0.0001 and a 95% CI of 4.74 to 91.2) between vaccinated and unvaccinated children, respectively.[93] In addition, the First Mawson

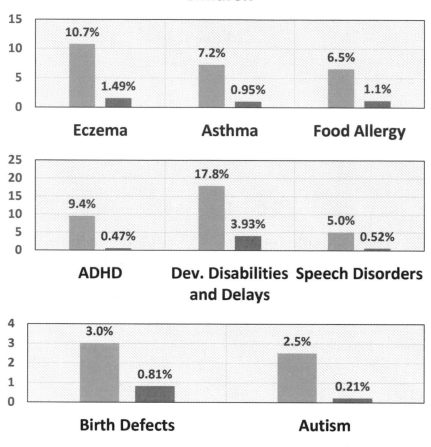

Figure 2.12—Comparison of percentage of US children with eczema,[83] asthma,[84] food allergy,[85] ADHD,[86] developmental disabilities and delays,[87] speech disorders,[88] birth defects,[89] and autism[90] among vaccinated versus unvaccinated (Statistical Evaluation of Health Outcomes of the Unvaccinated, Joy Garner, The Control Group, February 9, 2021, thecontrolgroup.org).

Study reported an odds ratio of 4.2 (p-value = 0.013 and a 95% CI of 1.2 to 14.5) between vaccinated and unvaccinated children for both autism and ADHD, based on his survey of homeschooled students.[94]

Figure 2.13 shows the results from the study "The Relationship between Vaccine Refusal and Self-Report of Atopic Disease in Children," published in the *Journal of Allergy and Clinical Immunology* in 2005.[95] The lead author, Dr. Rachel Enriquez, was affiliated with the Division of Allergy, Pulmonary and Critical Care Medicine, at Vanderbilt University in Nashville, Tennessee. The authors found relative risks of 11.4 (p-value < 0.0001) and 10 (p-value = 0.0002) for asthma and hay fever, respectively, when investigating parental reports of atopy or common allergies among vaccinated and unvaccinated children in the US.[96] The study cohort included 515 unvaccinated, 423 partially vaccinated, and 239 fully vaccinated children. Previously discussed studies by Mawson,[97] Hooker and Miller,[98] and the unpublished Netherlands study[99] affirm the results of this study.[100]

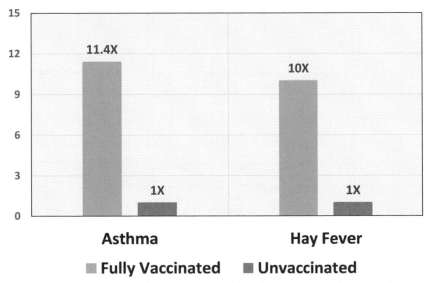

The Relationship between Vaccine Refusal and Self-Report of Atopic Disease in Children

Relative Risk of Asthma and Hay Fever in Vaccinated and Unvaccinated Children

Figure 2.13—Relative risk of asthma and hay fever reported from vaccinated versus unvaccinated children in the United States (Enriquez et al. 2005).

Association between Aluminum Exposure from Vaccines before Age 24 Months and Persistent Asthma at Age 24 to 59 Months

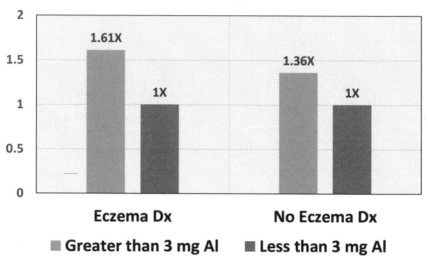

Incidence of Persistent Asthma Diagnoses in Children Exposed to Aluminum in Vaccines Prior to Age 24 Months

Figure 2.14—Incidence of persistent asthma in children aged 24 to 59 months based on aluminum exposure from vaccines given prior to age 24 months. Children with eczema diagnoses are considered separately from those without eczema diagnoses (Daley et al. 2022).

Figure 2.14 shows the results from the study "Association between Aluminum Exposure from Vaccines before Age 24 Months and Persistent Asthma at Age 24 to 59 Months," published in *Academic Pediatrics* in 2022.[101] The lead author, Dr. Matthew Daley, is from the Institute of Health Research, Kaiser Permanente Colorado, in Aurora. As a Kaiser researcher, Dr. Daly has access to CDC's Vaccine Safety Datalink (VSD) and studied a cohort of 326,991 children from the VSD. Aluminum exposure from vaccines given before 24 months of age was totaled for each child. Children exposed to over 3 milligrams of aluminum were 36% more likely (95% CI of 1.21 to 1.53) to receive a persistent asthma diagnosis between 24 and 59 months of age.[102] Similarly, children diagnosed with eczema exposed to greater than 3 milligrams of aluminum were 61% more likely (95% CI of 1.04 to 2.48) to be diagnosed with persistent asthma.[103]

Summary

	Mawson et al. 2017	Hooker and Miller 2020	Hooker and Miller 2021	Lyons-Weiler and Thomas 2021	Dutch Survey 2004	Control Group 2021	Enríquez et al. 2005	Daley et al. 2022
ADD and/or ADHD	✓		✓			✓		
Allergies	✓		✓		✓			
Asthma		✓	✓	✓	✓	✓	✓	✓
Autism	✓		✓			✓		
DD, LD or NDD	✓	✓				✓		
Eczema	✓			✓	✓			
Ear Infections	✓	✓	✓	✓	✓			
Gastro. Disorders		✓	✓	✓				
Respiratory Infections				✓				
Seizures					✓			

Table 2.1—Summary of results in comparing health outcomes of vaccinated versus unvaccinated children. Significantly higher odds ratios, relative risks, or incidences are denoted by a ✓.

Asthma was the most prevalent diagnosis related to the vaccination schedule, where vaccinated children had a higher incidence than unvaccinated children in seven separate studies covered in this chapter.[104, 105, 106, 107, 108, 109, 110] Respiratory infections[111] and seizures[112] showed an association with the vaccination schedule in one study each. However, the other research studies included in this chapter did not specifically consider these diagnoses. For example, although not included in the figures or Table 2.1, Lyons-Weiler and Dr. Thomas remarked that 5.3% of the vaccinated children in their study were diagnosed with ADHD compared to none of the unvaccinated children.[113] Also, the authors of the Dutch study commented that eight of their vaccinated subjects were diagnosed with autism compared to none of their unvaccinated subjects.[114]

CHAPTER 3

Thimerosal in Vaccines

Thimerosal is one of the most questionable components of vaccines. Unfortunately, some vaccines distributed in the United States still contain thimerosal. It is a chemical compound containing almost 50% mercury by mass and is used as a preservative in vaccines, primarily those formulated in multidose vials, to prevent microbial contamination. Many authors and researchers have written about thimerosal and the controversy surrounding it. In 2006, David Kirby wrote the book *Evidence of Harm,*[1] and more recently, in 2015, Robert F. Kennedy Jr. wrote *Thimerosal: Let the Science Speak.*[2] Eric Gladen featured the subject in his 2014 documentary film *Trace Amounts.*[3]

It seems counterintuitive to inject mercury, one of the most toxic elements on Earth, directly into your body, yet we have done this since the invention of thimerosal in the 1920s.[4] Unfortunately, no one has shown this organomercury compound to be safe; to the contrary, many countries across the globe have banned its use.[5]

US government officials recognized the problem with thimerosal in 1999 when Dr. Neal Halsey, a vaccinologist from Johns Hopkins University, completed a simple calculation.[6] He added the cumulative mercury level in the CDC's recommended infant vaccination schedule at the time and found that the total dosage far exceeded safety limits set by the FDA and the EPA.[7] For an infant to tolerate a single mercury-laced injection without harm based on these guidelines, he or she would have to weigh over 200 kg (around 440 pounds).[8] Because of Halsey's calculation, Department of Health and Human Services (DHHS) officials sent a flurry of emails leading the CDC to commission a study to determine whether mercury in the ever-increasing childhood vaccine schedule could be causing the increase of autism and other neurodevelopmental disorders.

Figure 3.1 shows the results from the abstract for the presentation "Increased Risk of Developmental Neurological Impairment after High Exposure to Thimerosal-Containing Vaccines in First Month of Life," published for the CDC's Epidemic Intelligence Service meeting in 1999 (Verstraeten Study).[9] Vaccine safety advocates from SafeMinds obtained the CDC abstract through the Freedom of Information Act (FOIA). Dr. Thomas Verstraeten, a Dutch epidemiologist recruited by the CDC's Immunization Safety Office from the Epidemic Intelligence Service fellowship program, was the study's lead author.[10] He examined the outcomes of babies given the thimerosal-containing hepatitis B vaccine two weeks after birth, as well as thimerosal-containing hepatitis B immune globulin given to infants whose mothers had the hepatitis B virus.[11] He found stunning and alarming results. For example, infants exposed to the highest possible levels of thimerosal during their first month of life (greater than 25 micrograms of mercury) had a 7.6 times higher risk of autism diagnosis (95% CI of 1.8 to 31.5) than their unexposed peers.[12] Additional results indicated that these infants also carried a 1.8 times higher risk for neurodevelopmental disorders (95% CI of 1.1 to 2.8), 5.0 times higher risk for nonorganic sleep disorders (95% CI of 1.6 to 15.9), and 2.1 times higher risk for speech disorders (95% CI of 1.1 to 4.0) compared to the zero-thimerosal exposure group.[13]

Increased Risk of Developmental Neurological Impairment
after High Exposure to Thimerosal-Containing Vaccines
in First Month of Life

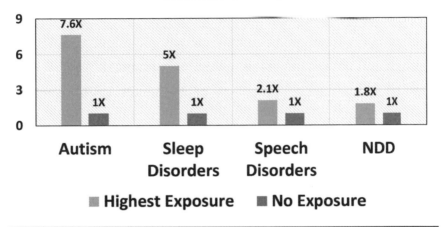

Figure 3.1—Relative risk of autism, sleep disorders, speech disorders, and
neurodevelopmental disorders (NDD) in those children exposed to the highest
levels of thimerosal in hepatitis B vaccines and immune globulin versus those
with no thimerosal exposure within the first month of life.

This discovery sent shock waves through the DHHS, including the
FDA and CDC. Government officials convened a secret meeting in
2000 at the Simpsonwood Retreat Center in Norcross, Georgia (away
from CDC Headquarters in Atlanta to keep records of the meeting
hidden from the public).[14] Government officials, university experts,
and industry representatives attended the meeting and discussed how
to hide this information from the public.[15] They determined strategies
to statistically dilute the relationship between thimerosal and autism,

among other disorders, and Verstraeten and his coresearchers at the CDC quickly executed these alterations.[16]

Five study iterations later, the CDC had massaged Verstraeten's data to the point that the strong relationship between thimerosal exposure and neurodevelopmental disease had vanished.[17] Before most of the manipulation of his reports was finalized, Verstraeten fled the CDC for an overseas position with vaccine giant GlaxoSmithKline. He had very little input on the 2003 research paper that bore his name.[18] When the journal *Pediatrics* published the paper, the CDC trumpeted loudly that it meant thimerosal did not cause autism, despite protests from Verstraeten. In 2004, Verstraeten wrote a letter to *Pediatrics*, stating that the study was "neutral" and could not rule out such a relationship.[19]

However, DHHS had hatched all this as a plan to indemnify thimerosal to avoid paying out mounting injury claims to the National Vaccine Injury Compensation Program (NVICP) for petitioners who had received autism diagnoses, likely due to thimerosal exposure. CDC researchers and other officials skillfully executed the plan, culminating in May of 2004, when the prestigious Institute of Medicine's (IOM) Immunization Safety Review Committee declared that thimerosal was not causal in autism, basing its opinion on five separate trumped-up epidemiology studies.[20]

Peer-reviewed, reputable scientific literature has repeatedly documented the harm caused by thimerosal-containing vaccines. Hooker and his coauthors uncovered the CDC's dubious methods in hiding the toxic effects of thimerosal in their paper "Methodological Issues and Evidence of Malfeasance in Research Purporting to Show Thimerosal in Vaccines is Safe" in the journal *BioMed Research International* in 2014.[21] Using the FOIA and independent analysis of the data, the authors of this paper revealed fatal flaws in each of the five epidemiologic studies used by the IOM to exonerate thimerosal in the autism epidemic. Authors of the five flawed studies resorted to hiding data from the public to eliminate downward trends in autism rates associated with removing thimerosal from vaccines.[22] In some cases, the authors repeatedly analyzed data using different inclusion criteria, such as counting autism cases in children down to birth, who though vaccinated, were much too young to have received an autism diagnosis.[23]

Researchers performed these and other flawed analyses to obfuscate significant relationships.[24] Scientists in several of the papers committed "overmatching" errors where children in "vaccinated" and "control" groups were too closely matched to one another to make a valid comparison.[25] Instead of comparing children who received no thimerosal to exposed groups of children, the authors juxtaposed children receiving some thimerosal against those who received a small additional increment and calculated the risk associated with that increment.[26]

Figure 3.2 shows results from the paper "Thimerosal Exposure in Infants and Neurodevelopmental Disorders," published in the *Journal of Neurological Sciences* in 2008.[27] Dr. Heather Young, the lead author, is a professor and epidemiologist at the George Washington University School of Public Health and Health Services.[28] Coauthors on the paper, Dr. Mark Geier and his son, David Geier, played key roles in the debate surrounding thimerosal-containing vaccines. Dr. Mark Geier is a former NIH scientist and a physician who, with his son, completed a flurry of studies beginning in the early 2000s that disclosed the harm caused by thimerosal in vaccines.[29, 30, 31, 32, 33] Due to their tenacity and the assistance of Rep. Dave Weldon (R-FL) and Rep. Dan Burton (R-IN), the CDC granted these researchers access to the Vaccine Safety Datalink (VSD), the same database used to produce the fatally flawed Verstraeten Study.[34] In their first VSD paper, Dr. Young and the Geiers found that a difference of 100 micrograms of mercury from thimerosal in infant vaccines given within the first seven months of life was associated with a 2.87 times greater rate of autism (p-value < 0.05 and a 95% CI of 1.19 to 6.94), 2.44 times greater rate of autism spectrum

Thimerosal Exposure in Infants and Neurodevelopmental Disorders

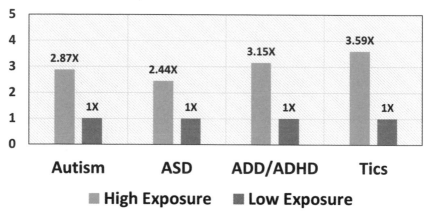

Rate Ratios of Diagnosis For Children with High and Low Thimerosal Exposure in their Infant Vaccines

Figure 3.2—Increase in risk of autism, ASD, ADD/ADHD, and tics with an increased 100 micrograms between high and low exposure groups for mercury exposure by seven months of age (Young et al. 2008).

disorder (p-value < 0.05 and a 95% CI of 1.16 to 5.10), 3.15 times greater rate of ADD/ADHD (p-value < 0.001 and a 95% CI of 2.38 to 4.17), and 3.59 times greater rate of tics (p-value < 0.001 and a 95% CI of 1.64 to 6.79).[35] Dr. Young and her coauthors used a statistical metric called "rate ratio," similar to the odds ratio.[36] Instead of comparing the odds of diagnosis in each group of children, however, the rate ratio compares the incidence, or "rate of diagnosis," in the high-exposure group to the incidence in the low-exposure group.

Figure 3.3 shows results from the paper "A Two-Phase Study Evaluating the Relationship between Thimerosal-Containing Vaccine Administration and the Risk for an Autism Spectrum Disorder Diagnosis in the United States," published in the journal *Translational Neurodegeneration* in 2013.[37] In this follow-up investigation of the VSD,

Thimerosal-Containing Hepatitis B Vaccine and Autism

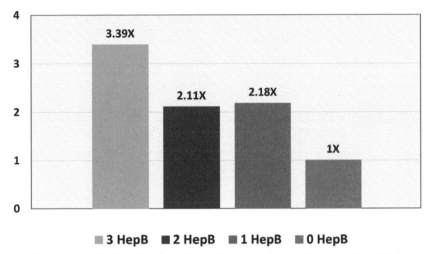

Odds Ratios for an Autism Diagnosis from Thimerosal-Containing Hepatitis B Vaccines

Figure 3.3—Odds ratios for autism diagnosis due to thimerosal-containing hepatitis B vaccines versus thimerosal-free hepatitis B vaccines (Geier et al. 2013).

the Geiers discovered comparable results for autism and thimerosal via exposure to the hepatitis B vaccine.[38] Children receiving one vaccination within the first month of life showed 2.18 times greater odds of autism (p-value < 0.00001 and a 95% CI of 1.74 to 2.73).[39] Children receiving two vaccines within the first two months of life showed 2.11 times greater odds of autism (p-value < 0.0001 with a 95% CI of 1.68 to 2.64).[40] Finally, children receiving the entire three-shot series within the first six months of life showed 3.39 times greater odds of autism (p-value < 0.001 and a 95% CI of 1.60 to 7.18).[41]

Figure 3.4 shows information from two papers, "Thimerosal Containing Hepatitis B Vaccination and the Risk of Diagnosed Specific Delays in Development in the United States: A Case-Control Study in the Vaccine Safety Datalink," published in the *North American Journal*

Thimerosal-Containing Hepatitis B Vaccine, Tics, and Specific Delays in Development

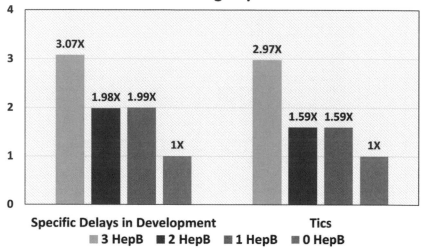

Figure 3.4—Odds ratios for specific delays in development and tics diagnoses due to thimerosal-containing hepatitis B vaccines versus thimerosal-free hepatitis B vaccines (Geier et al. 2014, Geier et al. 2015).

of Medical Sciences in 2014,[42] and "Thimerosal Exposure and Increased Risk of Diagnosed Tic Disorder in the United States: A Case-Control Study," published in the journal *Interdisciplinary Toxicology* in 2015.[43] A team of scientists led by the Geiers completed the research.[44, 45] Practitioners diagnosed children with specific delays in development from the hepatitis B vaccine at 1.99 times greater odds in the exposed group for one vaccine within the first month of life (p-value < 0.0001 and a 95% CI of 1.89 to 2.11), 1.98 times greater odds in the exposed group for two vaccines in the first two months of life (p-value < 0.0001 and a 95% CI of 1.87 to 2.09), and 3.07 times greater odds if infants received all three vaccines in the series within the first six months of life (p-value < 0.00001 and a 95% CI of 2.50 to 3.77).[46] Practitioners also diagnosed children with tics at 1.59 times greater odds in those who received the first shot by one month and two shots by two months (p-value < 0.00001 and a 95% CI of 1.29 to 1.98) and 2.97 times greater odds in those receiving all three vaccines by six months (p-value < 0.005 and a 95% CI of 1.46 to 6.05).[47] The control group in these studies received thimerosal-free hepatitis B vaccines.[48, 49]

Figure 3.5 shows results from the papers "Thimerosal Exposure and Disturbance of Emotions Specific to Childhood and Adolescence: A Case-Control Study in the Vaccine Safety Datalink (VSD) Database," published in the journal *Brain Injury* in 2017,[50] and "Premature Puberty and Thimerosal Containing Hepatitis B Vaccination: A Case-Control Study in the Vaccine Safety Datalink," published in the journal *Toxics*

Thimerosal-Containing Hepatitis B Vaccines, Emotional Disturbances, and Premature Puberty

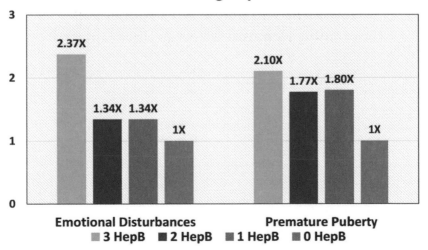

Figure 3.5—Odds ratios for emotional disturbances and premature puberty diagnoses due to thimerosal-containing hepatitis B vaccines versus thimerosal-free hepatitis B vaccines (Geier et al. 2017, Geier et al. 2018).

in 2018.[51] Practitioners diagnosed children with emotional disturbances[52] and premature puberty[53] with greater odds when exposed to
thimerosal-containing hepatitis B vaccines.[54] Interestingly, for all diagnoses that the Geiers considered when using the VSD, odds ratios in
the one-thimerosal-containing-vaccine in the first month of life and
the two-thimerosal-containing-vaccines in the first two months of
life analyses were very similar and did not increase significantly with
increasing exposure.[55, 56] This may be due to the "healthy user bias,"
where healthy subjects continue to receive vaccines, but subjects with
health issues limit or curtail further vaccination.[57] However, odds ratios
consistently increased for the highest exposure level, three thimerosal-containing hepatitis B vaccines by six months of life.[58]

Figure 3.6 shows results from the study "Hepatitis B Vaccination in
Male Neonates and Autism Diagnosis, NHIS 1997–2002," published

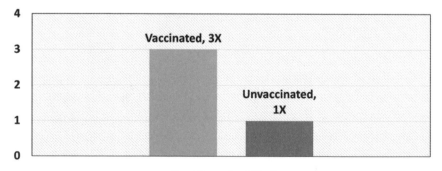

Hepatitis B Vaccination in Male Neonates and Autism Diagnosis, NHIS 1997–2002

Relative Risk of Autism Diagnosis in Male Newborns Vaccinated with Thimerosal-Containing Hepatitis B versus Unvaccinated

Autism in Males

Figure 3.6—Relative risk of autism diagnosis in males vaccinated with
thimerosal-containing hepatitis B vaccine during the first month of life versus
unvaccinated control males (Gallagher and Goodman 2010).

in the *Journal of Toxicology and Environmental Health Part A* in 2010.[59] Dr. Carolyn Gallagher, the lead author, completed this research while finishing her PhD program in Population Health and Clinical Outcomes Research at the Center for Public Health and Health Policy Research at State University of New York, Stony Brook.[60] Gallagher and her coinvestigator, Dr. Melody Goodman, studied the neonatal thimerosal-containing hepatitis B vaccine. They found that boys receiving this shot within the first month of life were three times as likely to receive an autism diagnosis compared to those who delayed getting the vaccine until after the first month of life.[61] Nonwhite children bore a greater risk.[62]

Figure 3.7 shows results from the study "Hepatitis B Triple Series Vaccine and Developmental Disability in US Children Aged 1–9

Hepatitis B Triple Series Vaccine and Developmental Disability in US Children Aged 1–9 Years

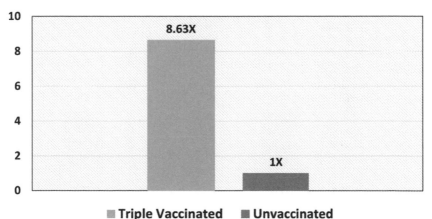

Figure 3.7—Odds ratios for special education services in males vaccinated with thimerosal-containing hepatitis B vaccine series versus unvaccinated control males (Gallagher and Goodman 2008).

Years," published in *Toxicology & Environmental Chemistry* in 2008.[63] In this study, Gallagher and Goodman found that boys receiving the thimerosal-containing three-shot hepatitis B series were nearly nine times more likely to receive special education services compared to boys who skipped the hepatitis B vaccine altogether.[64] Their analysis also affirms the previous connection Verstraeten found in 1999 between early thimerosal exposure and autism, as well as other developmental, language, and speech delays, ADD, and tics.[65, 66]

Figure 3.8 shows results from the paper "Early Thimerosal Exposure and Psychological Outcomes at 7 to 10 Years," published in the *New England Journal of Medicine* in 2007.[67] Dr. William Thompson, the senior epidemiologist in the Influenza Division and formerly from

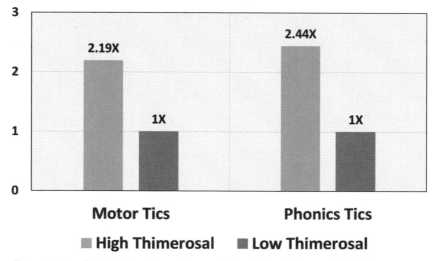

Figure 3.8—Odds ratios for motor and phonic tics in boys with high exposure versus low thimerosal exposure in infant vaccines (Thompson et al. 2007).

the Immunization Safety Office at the CDC, Atlanta, Georgia, is the lead author.[68] Although the CDC continues to claim that thimerosal is perfectly safe, its own studies show definitively that thimerosal exposure causes tics. Using data from the CDC's VSD, Thompson also demonstrated that boys receiving higher levels of thimerosal via infant vaccines over the first seven months of life had 2.19 times greater odds of having motor tics (p-value < 0.05 and a 95% CI of 1.02 to 4.67) and 2.44 times greater odds of phonic tics compared to those boys receiving lower levels of thimerosal (p-value < 0.05 and a 95% CI of 1.12 to 5.35).[69]

Unlike previously discussed research, CDC study authors did not include a "zero exposure" control in this study.[70] Instead, they specified "high" and "low" exposure groups where the difference in thimerosal exposure was two standard deviations based on the cumulative exposure in the males in the cohort between birth and seven months of age.[71] Boys in the study had a median level of exposure of 112.5 micrograms of mercury, and less than 2% of the cohort had no thimerosal exposure.[72] By narrowing the gap between high- and low-exposure groups, the CDC study authors biased the study to hide a relationship between tics and thimerosal exposure.[73] In a subsequent CDC publication with John Barile from Georgia State University, Thompson affirmed the relationship between thimerosal and tics.[74]

Figure 3.9 shows results from the paper "Thimerosal Exposure in Infants and Developmental Disorders: A Retrospective Cohort Study in the United Kingdom Does Not Support a Causal Association," published in *Pediatrics*.[75] Nick Andrews, the lead author, is an epidemiologist from the Statistics Unit and Immunisation Department, Health Protection Agency at the Communicable Disease Surveillance Center in London, UK.[76] Like Thompson,[77] Andrews and his coinvestigators reported a consistent relationship for tics among children receiving thimerosal-containing DTP/DT vaccines at three and four

Thimerosal Exposure in Infants and Developmental Disorders:
A Retrospective Cohort Study in the United Kingdom Does Not
Support a Causal Association

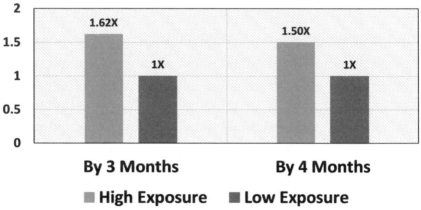

Hazard Ratios for Tics in Children per Dose of Thimerosal-Containing DTP/DT Vaccine

Figure 3.9—Hazard ratios for tics in children with high versus low thimerosal exposure in infant vaccines (Andrews et al. 2004).

months of age in the UK.[78] Hazard ratios shown in Figure 3.9 reflect the increased risk of tic disorder by taking one additional thimerosal-containing DTP/DT vaccine by three months of age or by four months of age.[79] Children following the UK vaccine schedule receive four DTP/DT vaccine doses before their first birthday and as many as three doses before three months of age.[80] All children in this analysis received at least three DTP/DT vaccines prior to their first birthday.[81] This shows that children exposed to greater levels of thimerosal earlier in infancy have a higher incidence of tics.[82] Curiously, the results from this study do not support the paper's title.

While stubbornly refusing to publicly admit thimerosal's toxicity, between 2001 and 2004, the CDC recommended a phase-down of the production of thimerosal-containing infant vaccines, including the hepatitis B, *Haemophilus influenzae* B, and DTaP vaccines.[83] Contrary to what the CDC has led the populace to believe, they never removed mercury from the vaccine schedule but instead redistributed it insidiously. As manufacturers phased down mercury from existing childhood vaccines, the CDC added annual flu shots, many of which still contain thimerosal,[84] for infants as young as six months of age and children each year afterward. In this fashion, children may be exposed to an additional 25 micrograms of mercury every year of their lives.

Although the CDC recommends avoiding mercury exposure during pregnancy, the agency now simultaneously endorses the use of thimerosal-laced flu shots, given during any trimester of pregnancy, to the same population.[85] Their claims of its safety during pregnancy are unfounded, given that the FDA has never approved flu shots for such use. Instead, open literature shows that the opposite is true. Package inserts for flu vaccines typically have disclaimers regarding use in pregnant women. For example, the Fluvirin® (Seqirus, Inc.) package insert specifically states, "Safety and effectiveness of Fluvirin® have not been established in pregnant women . . ."[86] We discuss studies regarding thimerosal exposure in pregnancy further in Chapter 11.

The CDC website regarding the 2022–23 flu vaccine supply claimed that 93% of flu shots were thimerosal-free.[87] However, it is unclear whether this means that 93% of all flu vaccines were thimerosal-free or 93% of all vials containing the flu vaccines (including multidose vials) were thimerosal-free, whereas thimerosal-containing flu vaccines consist of ten doses of the vaccine per vial. If the latter is true, only 57% of all vaccines in the 2022–23 flu shot season were thimerosal-free.

While this lack of transparency is troubling for US consumers, even worse is the fact that thimerosal-containing versions of childhood vaccines are still used in developing countries. According to the Pan

American Health Organization's Minamata Treaty website, thimerosal-containing vaccines are used to inoculate over 80 million children worldwide.[88] The same website falsely states that thimerosal is not linked to neurodevelopmental disorders.

Summary

	Verstraeten et al. 1999	Young et al. 2008	Geier et al. 2013-2018	Gallagher et al. 2010	Gallagher et al. 2008	Thompson et al. 2007	Andrews et al. 2004
ADD/ADHD		✓					
Autism	✓	✓	✓	✓			
ASD		✓					
Emotional Disturbances			✓				
NDD	✓						
Premature Puberty			✓				
Sleep Disorders	✓						
Specific Delays in Dev./Special Education				✓	✓		
Speech Disorders	✓						
Tics		✓	✓			✓	✓

Table 3.1—Summary of results comparing health outcomes of children exposed to thimerosal-containing vaccines. Significantly higher odds ratios, relative risks, hazard ratios, or incidences are denoted by a ✓.

Autism is significantly related to thimerosal exposure in four of the studies highlighted in this chapter.[89, 90, 91, 92, 93, 94, 95] Autism spectrum disorder (ASD), considered a separate diagnosis from autism, is also related to thimerosal exposure in the study by Young.[96] Thimerosal exposure correlated with tics in four of the studies from this chapter.[97, 98, 99, 100] Thompson further differentiates between motor tics and phonic tics, which were both related to thimerosal exposure in boys in the study.[101]

Specific delays in development and special education services (SPED) showed significant relationships in two studies.[102, 103, 104,105, 106] The Geier studies between 2013 and 2018 are displayed in a single column, as each study considered a single disorder separately.[107, 108,109, 110]

CHAPTER 4

Live Virus Vaccines: MMR, Polio, and Rotavirus

The MMR vaccine is the tip of the spear regarding the modern debate around vaccine safety. Dr. Andrew Wakefield and eleven colleagues at the Royal Free Hospital in London published their findings that eight out of the 12 cases of autistic enterocolitis they had seen occurred after patients received the MMR vaccine.[1] To be clear, Wakefield and his coauthors did not state in this research article originally published in *Lancet* that the MMR caused autism or autistic enterocolitis. They merely pointed out the timing of the vaccine before the onset of symptoms. The MMR vaccine was not the focus of the paper. However, the brief section that mentioned the vaccine set off World War III with the pharmaceutical industry focusing all its weapons squarely on Dr. Wakefield. In his 2010 book, *Callous Disregard*,[2] Dr. Wakefield details the controversial events that followed. Rather than covering those here, we highlight vax-unvax literature that focuses on live virus vaccines, including the MMR, polio, and rotavirus vaccines and their associated outcomes.

Figure 4.1 shows the results of the paper "Age at First Measles-Mumps-Rubella Vaccination in Children with Autism and School-Matched

Age at First Measles-Mumps-Rubella Vaccination in Children
with Autism and School-Matched Control Subjects:
A Population-Based Study in Metropolitan Atlanta

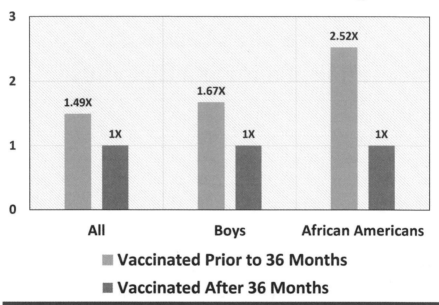

Odds of Autism with MMR Vaccine
Before and After 36 Months of Age

Figure 4.1—Odd ratios for autism for different groups of children, comparing
children vaccinated before 36 months of age to those vaccinated after 36
months of age (DeStefano et al. 2004).

Control Subjects: A Population-Based Study in Metropolitan Atlanta,"
published in the journal *Pediatrics* in 2004.[3] Dr. Frank DeStefano, the
former director of the CDC's Immunization Safety Office, is the paper's
lead author. The study investigators report 1.49 times greater odds of
an autism diagnosis in children receiving the MMR vaccine before
36 months of age versus children receiving the MMR vaccine after
36 months of age (95% CI of 1.04 to 2.14).[4] Boys receiving the MMR
prior to 36 months of age possessed 1.67 greater odds for an autism

diagnosis compared to boys receiving the MMR after 36 months of age (95% CI of 1.10 to 2.53).[5] DeStefano and his coauthors dismissed these statistically significant findings as artifacts of vaccine requirements for early intervention special education services. However, if early vaccine requirements were responsible for the results, girls would also show a significant relationship between MMR timing and autism incidence. They did not. Instead, girls vaccinated before 36 months of age showed an odds ratio of 1.06 with a 95% confidence interval of 0.51 to 2.20 when compared to girls who delayed the vaccine until after 36 months of age.[6]

Results not published, obtained from CDC senior scientist Dr. William Thompson, show that African American children receiving the MMR prior to 36 months had 2.4 times greater odds of autism diagnosis compared to children receiving the MMR after 36 months. This finding was statistically significant (95% CI is 1.4 to 4.4). However, rather than publishing these results, the CDC scientists removed all African American children from the sample who did not possess a valid State of Georgia birth certificate. They did this to obviate the statistically significant finding and instead reported no difference in autism incidence between the two groups of children.

Figure 4.2 shows results from the paper "Reanalysis of CDC Data on Autism Incidence and Time of First MMR Vaccination," published in the *Journal of American Physicians and Surgeons* in 2018.[7] Dr. Brian Hooker is the author of this publication. African American boys receiving the MMR vaccine prior to 36 months of age possessed 3.86 times greater odds of an autism diagnosis compared to African American boys receiving the MMR vaccine after 36 months of age (p-value = 0.005 and a 95% CI of 1.49 to 10.0).[8] Dr. Hooker obtained these results using the dataset from the DeStefano publication.[9] DeStefano and his coauthors did not complete an analysis specific to African American boys in their original publication.

Reanalysis of CDC Data on Autism Incidence and Time of First MMR Vaccination

Odds of Autism with MMR Vaccine Before and After 36 Months of Age

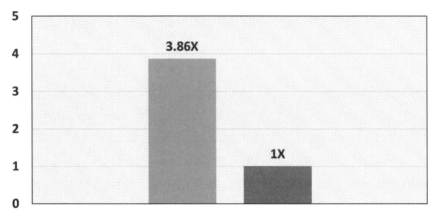

African American Boys

■ **Vaccinated Prior to 36 Months**

■ **Vaccinated After 36 Months**

Figure 4.2—Odd ratios for autism, in African American boys, comparing children vaccinated prior to 36 months of age to children vaccinated after 36 months of age (Hooker 2018).

Figure 4.3 shows more results from Hooker's paper "Reanalysis of CDC Data on Autism Incidence and Time of First MMR Vaccination," published in the *Journal of American Physicians and Surgeons* in 2018. Children receiving the MMR vaccine prior to 36 months of age had 2.52 times greater odds of a diagnosis of "autism without mental retardation" compared to children receiving the MMR vaccine after 36 months of age (p-value = 0.012 and a 95% CI of 1.23 to 5.17).[10] (Mental retardation is defined as an IQ of 70 or below.) Autism without mental retardation was termed "isolated autism" within the analysis of DeStefano et al.[11] This result was also obtained by Dr. William Thompson of the CDC but was omitted from the CDC's final published study.

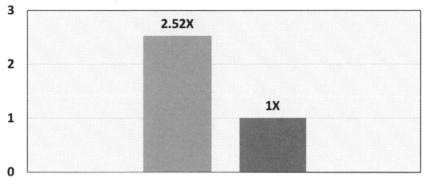

Odds of Autism without Mental Retardation for MMR Vaccine Before and After 36 Months of Age

All Children

■ **Vaccinated Prior to 36 Months**
■ **Vaccinated After 36 Months**

Figure 4.3—Odd ratios for autism without mental retardation or "isolated autism," comparing children vaccinated prior to 36 months of age to children vaccinated after 36 months of age (Hooker 2018).

Figure 4.4 shows results from the paper "Is Measles Vaccination a Risk for Inflammatory Bowel Disease?" published in the journal *Lancet* in 1995.[12] Nick P. Thompson, from the Royal Free Hospital School of Medicine in London, UK, is the study's lead author. Dr. Wakefield is the corresponding author of the study. Recipients of the live measles vaccine had a 3.01 times greater relative risk of a diagnosis of Crohn's disease (p-value = 0.004 and a 95% CI of 1.45 to 6.23) and a 2.53 times greater relative risk of ulcerative colitis compared to unvaccinated individuals (p-value = 0.03 and a 95% CI of 1.15 to 5.58).[13] The vaccinated cohort was from a 1964 randomized trial of measles vaccine recipients and consisted of 3,545 individuals who responded to a follow-up survey in 1994. The unvaccinated cohort was from the UK National Child

Is Measles Vaccination a Risk for Inflammatory Bowel Disease?

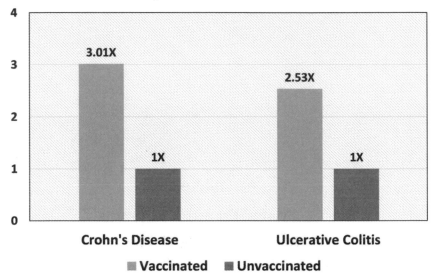

Figure 4.4—Relative risk of Crohn's disease and ulcerative colitis in children receiving the live measles vaccine versus those not receiving the live measles vaccine (Thompson et al. 1995).

Development Study of children born in 1958 and comprised 11,407 individuals surveyed in 1991.

Figure 4.5 shows results from the paper "Measles and Atopy in Guinea-Bissau," published in the journal *Lancet* in 1996.[14] Seif O. Shaheen, from the Centre for Primary Care and Public Health, Barts, and the London School of Medicine and Dentistry in London, UK, is the paper's lead author. The cohort consisted of 395 young adults from a semirural area in Guinea-Bissau. Thirty-three out of 129 vaccinated individuals were diagnosed with atopy compared to 17 out of 133 individuals who had measles infection, for an odds ratio of 2.8 (p-value = 0.01 and a 95% CI of 0.17 to 0.78).[15] The difference between the two groups was statistically significant. Atopy is a genetic predisposition to develop allergic diseases, including allergic rhinitis, asthma, and eczema.[16]

Measles and Atopy in Guinea-Bissau

Odds Ratios for Atopy in Vaccinated Children Versus Children Previously Infected with Measles

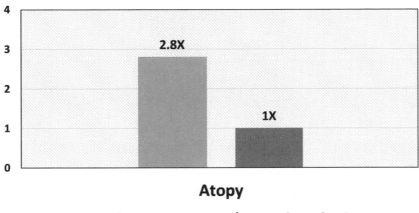

Atopy

■ **Measles Vaccination w/o Measles Infection**
■ **Measles Infection**

Figure 4.5—Odds ratios for atopy (allergies) in children receiving the measles vaccine versus children with measles infection (Shaheen et al. 1996).

Figure 4.6 shows results from the paper "Risk of Vaccine Induced Diabetes in Children with a Family History of Type 1 Diabetes," published in *The Open Pediatric Medicine Journal* in 2008.[17] Dr. John Barthelow Classen, chief executive officer of Classen Immunotherapies in Baltimore, Maryland, is the author of the paper. Among a cohort of all children born in Denmark between 1990 and 2000, those receiving all three recommended live virus, oral polio vaccines had an incidence of type 1 diabetes of 20.86 cases per 100,000 children as compared to children unvaccinated for polio, who had an incidence of type 1 diabetes of 8.27 cases per 100,000 children.[18] The difference in incidence between the two groups of children is statistically significant, with a rate ratio of 2.52 (95% CI of 2.06 to 3.08).[19] The oral polio vaccine was phased out in the US by 2000 and was replaced by the inactivated

Risk of Vaccine-Induced Diabetes in Children
with a Family History of Type 1 Diabetes

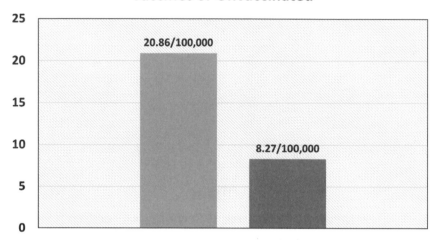

**Type I Diabetes Incidence per 100,000 Children
Vaccinated with All 3 Recommended Polio
Vaccines or Unvaccinated**

Figure 4.6—Incidence of Type 1 diabetes in children receiving all three recommended polio vaccines versus children unvaccinated for polio (Classen 2008).

polio vaccine. The oral polio vaccine is still distributed in other parts of the world.

Figure 4.7 shows results from the paper "Vaccination and Risk for Developing Inflammatory Bowel Disease: A Meta-Analysis of Case-Control and Cohort Studies," published in the journal *Clinical Gastroenterology and Hepatology* in 2015.[20] The study's lead author is Dr. Guillaume Pineton de Chambrun, affiliated with the Gastroenterology and Hepatology Department at Lille University Hospital in Lille, France. The study authors completed an analysis of three case-control studies with a total of 666 patients. Patients receiving poliomyelitis

Vaccination and Risk for Developing Inflammatory Bowel Disease: A Meta-Analysis of Case-Control and Cohort Studies

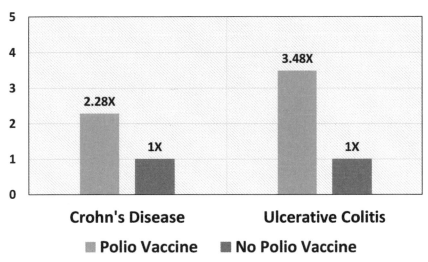

Relative Risk of Crohn's Disease and Ulcerative Colitis Following Polio Vaccination

Figure 4.7—Relative risk of Crohn's disease and ulcerative colitis in children vaccinated against polio versus children unvaccinated against polio (Pineton de Chambrun et al. 2015).

vaccines during childhood were 2.28 times more likely to have received a diagnosis of Crohn's disease (p-value < 0.05 and a 95% CI of 1.12 to 4.63) and 3.48 times more likely to receive a diagnosis of ulcerative colitis compared to their unvaccinated counterparts (p-value < 0.05 and a 95% CI of 1.2 to 9.71).[21] Both relationships were statistically significant.

Figure 4.8 shows results from the paper "Intussusception Risk and Health Benefits of Rotavirus Vaccination in Mexico and Brazil," published in the *New England Journal of Medicine* in 2011.[22] The lead author is Dr. Manish Patel, affiliated with the CDC in Atlanta, Georgia. Study authors state, regarding the effect observed for the Rotarix® vaccine, "An increased risk of intussusception 1 to 7 days

Intussusception Risk and Health Benefits of Rotavirus Vaccination in Mexico and Brazil

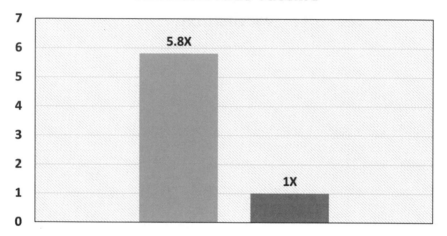

Figure 4.8—Odds ratio of intussusception in infants after receiving the first dose of Rotarix® rotavirus vaccine compared to unvaccinated controls (Patel et al. 2011).

after the first dose of RV1 (Rotarix®) was identified among infants in Mexico with the use of both the case-series method (incidence ratio, 5.3; 95% CI, 3.0 to 9.3) and the case-control method (odds ratio, 5.8; 95% CI, 2.6 to 13.0)."[23] According to the Children's Hospital of Philadelphia, "Intussusception is a life-threatening illness and occurs when a portion of the intestine folds like a telescope, with one segment slipping inside another."[24] It can lead to severe damage to the intestines, internal bleeding, and infection. If left untreated, this condition is fatal within two to five days.[25] GlaxoSmithKline manufactures Rotarix®.[26]

Figure 4.9 shows results from the paper "Risk of Intussusception Following Rotavirus Vaccination: An Evidence-Based Meta-Analysis

Risk of Intussusception Following Rotavirus Vaccination: An Evidence-Based Meta-Analysis of Cohort and Case-Control Studies

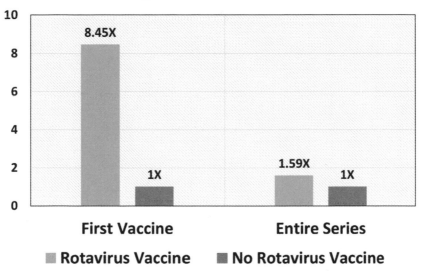

Figure 4.9—Odds ratios for intussusception following rotavirus vaccination compared to unvaccinated control infants (Kassim and Eslick 2017).

of Cohort and Case-Control Studies," published in the journal *Vaccine* in 2017.[27] Dr. Guy Eslick, affiliated with the Whitely-Martin Research Centre at the University of Sydney in Sydney, Australia, is the corresponding author. The study authors considered five separate case-control studies with a total of 9,643 children. They found an odds ratio of 8.45 (95% CI of 4.08 to 17.50) for intussusception following the first dose of the rotavirus vaccine as well as an odds ratio of 1.59 (95% CI of 1.11 to 2.27) for intussusception following all rotavirus vaccine doses compared to unvaccinated controls.[28] RotaTeq® (Merck)[29] and Rotarix® (GlaxoSmithKline)[30] were the only vaccines distributed within the studies considered in this meta-analysis.[31] The first rotavirus vaccine, Rotashield®, was withdrawn in the US because it induced a very high rate of intussusception.[32]

CHAPTER 5

Human Papillomavirus (HPV) Vaccines

Human papillomaviruses (HPV) encompass over one hundred strains of the virus that infect the skin cells, commonly known as epithelial cells.[1] HPV is ubiquitous and will infect nearly everyone with one or more strains at some point during their lifetime. While many strains of the virus show no apparent symptoms, others cause unsightly but harmless warts, also known as papilloma, that can erupt on the fingers, hands, feet, and genitals, while other strains, such as HPV 16 and HPV 18, have been associated with certain cancers, particularly cancer of the cervix. The overwhelming majority of HPV infections, even those involving cancer-associated strains, are self-limiting and resolve within two to three years, with the exception of a small subset of individuals in whom infections may persist and eventually evolve into cancerous lesions.[2] Fortunately, routine Pap testing can reliably identify precancerous cervical cells. In addition, practitioners can perform a loop electrosurgical excision, a simple, effective procedure to remove any suspect cells and virtually eliminate the risk of cancer.[3] Despite the baseline risk for cervical cancer being extremely low and a highly effective screening and treatment protocol for precancerous lesions, the pharmaceutical industry identified HPV as an opportunity to cash in by creating a vaccine that they can market as an anticancer inoculation.

In 2006, the FDA rushed Merck's Gardasil HPV vaccine through the approval process by way of the Prescription Drug User Fee Act.[4, 5] Enacted in 1992, the Prescription Drug User Fee Act allows drug companies to pay a substantial fee in return for the expedited approval of specified human drug and biological products.[6, 7] The original Gardasil vaccine contained antigens for four strains of HPV (6, 11, 16, and 18), two of which are primarily associated with cervical cancer and two of which are primarily associated with genital warts, as well as an aluminum adjuvant (amorphous aluminum hydroxyphosphate sulfate [AAHS]) to enhance immune response. During clinical trials, apart from a small subgroup of 300 patients, researchers did not test the Gardasil vaccine against a saline placebo. Instead, they gave the control group a solution containing this same aluminum adjuvant.[8] AAHS was a new adjuvant developed by Merck and first introduced in Europe with Procomvax, a vaccine against hepatitis B and *Haemophilus influenzae* B.[9] However, there are significant questions regarding the safety of AAHS, as researchers did not test it separately during the prelicensing evaluation of Procomvax.[10] Therefore, its use as a placebo during the Gardasil clinical trials was questionable and confounded the researchers' ability to determine the true safety profile of the vaccine. In addition, the placebo group was offered the vaccine six months into the clinical trial, meaning that no long-term follow-up was possible for either the safety or the efficacy of the vaccine.

In the original Gardasil clinical trial, both 2.3% of the vaccine group, 10,706 women, and 2.3% of the AAHS control group, 9,412 women, reported new conditions potentially indicative of an autoimmune disorder after receiving the vaccine or placebo.[11] Merck had already stacked the deck by using AAHS in the vaccine and the placebo groups. They were then able to dismiss the finding since the experimental and the placebo groups showed the same result.

Upon approval, Merck aggressively marketed Gardasil as a prophylactic against cervical cancer to females aged nine to twenty-six, and the FDA subsequently approved the vaccine for women up to forty-five years of age. Eventually, Merck also broadened its tent stakes and extensively promoted Gardasil for males between the ages of nine and forty-five.[12] Bolstered by Merck's untested claim that Gardasil also prevents anal cancer and various types of mouth and throat cancers,[13] the product was a boon for sales, with revenues in 2018 topping $3 billion.[14]

After the success of Merck's Gardasil, GlaxoSmithKline moved to enter the HPV vaccine market with their product, Cervarix, which the FDA approved in 2009.[15] Cervarix is formulated for protection against HPV 16 and 18, the strains predominantly associated with cervical cancer.[16] Like the Gardasil trials, clinical trials for Cervarix failed to test the vaccine against a true placebo. Instead, GlaxoSmithKline administered the aluminum hydroxide adjuvant-containing hepatitis A vaccine as the placebo.[17] This made it impossible to determine the actual safety profile of the new vaccine. Additionally, researchers never independently tested a component of the Cervarix adjuvant, monophosphoryl lipid A. The rate of new autoimmune conditions within the Cervarix trial was 0.8% in both the experimental and control groups.[18] As in the Gardasil vaccine trials, researchers dismissed any adverse events reported in the experimental group since there was no difference between the two groups.

In 2014, FDA approved Gardasil 9, which includes antigens for nine different strains of HPV and double the amount of AAHS adjuvant compared to the original Gardasil vaccine.[19] Within clinical trials leading up to the approval of the new vaccine, the control group was actually given the original Gardasil vaccine rather than a saline

Who Profits from Uncritical Acceptance of Biased Estimates of Vaccine Efficacy and Safety

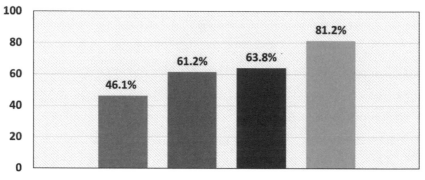

Age-Adjusted Rate of Adverse Events Due to HPV Vaccine Compared to All Vaccines 2006-2012

Figure 5.1—Age-adjusted rate of adverse events due to the HPV vaccine compared to all other vaccines, as reported in the CDC VAERS database (Tomljenovic and Shaw, 2012).

placebo.[20] Accordingly, 2.2% of the experimental group and 3.3% of the control group reported new medical conditions indicative of autoimmunity.[21] Despite these alarmingly high rates, Merck convinced FDA regulators to approve their product.

Since the rollout of HPV vaccines, savvy researchers have investigated the debatable nature of the clinical trials and FDA approval process. There is a large body of research regarding the extensive nature of HPV vaccine adverse events. This chapter specifically highlights vaccinated versus unvaccinated studies to seek out evidence concerning the safety and efficacy of these products.

Figure 5.1 shows results from the paper "Who Profits from Uncritical Acceptance of Biased Estimates of Vaccine Efficacy and Safety," written as a letter to the editor of the *American Journal of Public Health* in 2012.[22]

Drs. Lucija Tomljenovic and Chris Shaw, affiliated with the Neural Dynamics Research Group at the University of British Columbia in Vancouver, coauthored the letter. The 2012 Vaccine Adverse Event Reporting System (VAERS) data showed that more serious adverse reactions were attributed to Gardasil than all other vaccines, with Gardasil accounting for more than 60% of the total.[23] The Gardasil vaccine also accounted for 63.8% of all deaths, 61.2% of all life-threatening reactions, and 81.8% of all cases of permanent disability recorded in the CDC VAERS data.[24] While researchers cannot establish a causal relationship solely through passive reporting in VAERS, the disproportionate number of reports for Gardasil-associated events should signal further safety review.

Safety Concerns with Human Papilloma Virus Immunization in Japan: Analysis and Evaluation of Nagoya City's Surveillance Data for Adverse Events

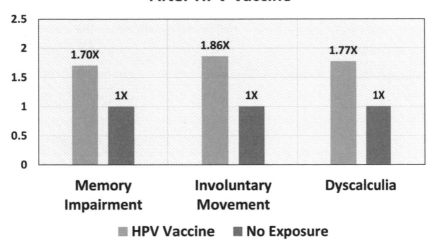

Figure 5.2—Odds ratios for neurological impairment, involuntary movement, and dyscalculia in 15- and 16-year-old women receiving the HPV vaccine versus unvaccinated controls (Yaju et al. 2019).

Figure 5.2 shows results from the paper "Safety Concerns with Human Papilloma Virus Immunization in Japan: Analysis and Evaluation of Nagoya City's Surveillance Data for Adverse Events," published in the *Japan Journal of Nursing Science* in 2019.[25] The lead author is Dr. Yukari Yaju, affiliated with the Department of Statistics in the Graduate School of Nursing Science at St. Luke's International University in Tokyo, Japan. Specifically, 15- and 16-year-old females possessed higher odds of memory impairment (95% CI of 1.24 to 2.33), involuntary movement (95% CI of 1.07 to 3.23), and dyscalculia (a learning disability in math) (95% CI of 1.00 to 3.13) in the group receiving the HPV vaccine versus those who did not receive it.[26] These relationships were statistically significant. The authors comment pointedly, "Based on our analysis using data from the Nagoya City surveillance survey, a possible association between HPV vaccination and distinct symptoms such as cognitive impairment or movement disorders exists."[27]

Figure 5.3 shows results from the paper "Behavioral Abnormalities in Female Mice following Administration of Aluminum Adjuvants and the Human Papillomavirus Vaccine Gardasil," published in *Immunology Research* in 2017.[28] The lead author, Dr. Rotem Inbar, is affiliated with the Zabludowicz Center for Autoimmune Diseases at the Sheba Medical Center and the Sackler Faculty of Medicine in Tel Aviv, Israel. Dr. Yehuda Shoenfeld, the incumbent of the Laura Schwarz-Kip Chair for Research of Autoimmune Diseases in the Sackler Faculty of Medicine at Tel Aviv University in Tel Aviv, Israel, is the corresponding author. Female mice receiving three human weight-equivalent doses of quadrivalent Gardasil vaccine produced anti-brain protein and anti-brain phospholipid antibody titers at 8.5 and 10 times that of unvaccinated control mice.[29] These antibody differences between Gardasil and control mice were statistically significant, with p-values less than 0.002.[30]

Behavioral Abnormalities in Female Mice following Administration of Aluminum Adjuvants and the Human Papillomavirus Vaccine Gardasil

Increases in Anti-Brain Protein and Anti-Brain Phospholipid Antibodies in Mice Due to HPV Vaccination

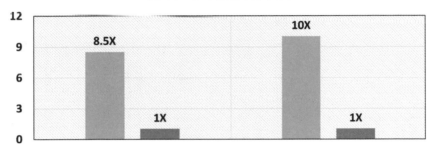

Figure 5.3—Increases in anti-brain (autoimmune) protein and phospholipid antibodies in mice receiving the HPV vaccine versus unvaccinated control mice (Inbar et al. 2017).

Figure 5.4 shows results from the paper "Human Papillomavirus Vaccination of Adult Women and Risk of Autoimmune and Neurological Diseases," published in the *Journal of Internal Medicine* in 2018.[31] The lead author is Dr. Anders Hviid, affiliated with the Department of Epidemiology Research at the Statens Serum Institut in Copenhagen, Denmark. In this study, a cohort of women from Sweden and Denmark showed a significantly higher risk of celiac disease after receiving the human papillomavirus vaccine than unvaccinated control women (95% CI of 1.29 to 1.89).

Human Papillomavirus Vaccination of Adult Women and Risk of Autoimmune and Neurological Diseases

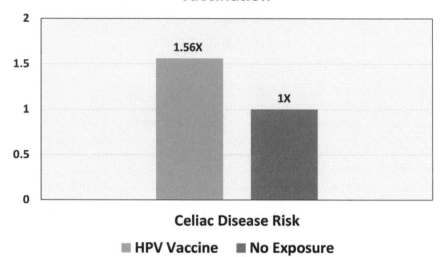

Figure 5.4—Risk of a celiac disease diagnosis after human papillomavirus vaccine versus unvaccinated controls (Hviid et al. 2018).

Figure 5.5 shows results from the paper "A Cross-Sectional Study of the Relationship between Human Papillomavirus Vaccine Exposure and the Incidence of Reported Asthma in the United States," published in the journal *SAGE Open Medicine* in 2019.[32] The study's lead author, David A. Geier, is affiliated with the Institute of Chronic Illnesses in Silver Spring, Maryland. Using data from the National Health and Nutrition Examination Survey, the study authors determined that HPV vaccine recipients have an 8.01 times greater incidence of asthma compared to persons not receiving the HPV vaccine (95% CI of 1.98 to 32.41).[33]

A Cross-Sectional Study of the Relationship between Human Papillomavirus Vaccine Exposure and the Incidence of Reported Asthma in the United States

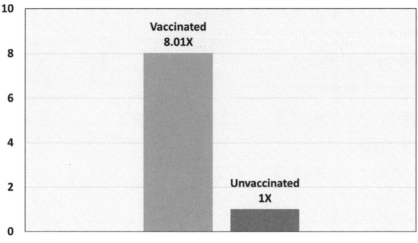

Figure 5.5—Odds ratio for asthma diagnosis after human papillomavirus vaccine versus unvaccinated controls (Geier et al. 2019).

Summary

Table 5.1 shows results from the five publications highlighted in Chapter 5.[34, 35, 36, 37, 38] In this case, each publication focused on a different set of vaccine sequelae. A large body of research highlights specific issues with the HPV vaccine. However, only the studies above compared vaccinated versus unvaccinated groups of individuals.

	Tomljenovic and Shaw 2012	Yaju et al. 2019	Inbar et al. 2017	Hviid et al. 2018	Geier et al. 2019
Serious Injury	✓				
Death	✓				
Permanent Disability	✓				
Memory Impairment		✓			
Involuntary Movement		✓			
Dyscalculia		✓			
Anti-Brain Protein Antibodies in Mice			✓		
Anti-Brain Phospholipid Antibodies in Mice			✓		
Celiac Disease				✓	
Asthma					✓

Table 5.1—Summary of results in comparing health outcomes of HPV-vaccinated versus unvaccinated individuals. Significantly higher odds ratios, relative risks, or incidences are denoted by a ✓.

CHAPTER 6

Vaccines and Gulf War Illness

In this chapter, we highlight scientific publications where Gulf War illness correlates with the number of vaccines received both prior to and during deployment. Many other publications are devoted to discussing issues related to different vaccines received during military service, particularly the anthrax vaccine. However, they did not include vaccinated versus unvaccinated comparisons.

Figure 6.1 shows results from the paper "Prevalence and Patterns of Gulf War Illness in Kansas Veterans: Association of Symptoms with Characteristics of Person, Place, and Time of Military Service," published in the *American Journal of Epidemiology* in 2000.[1] Dr. Lea Steele, affiliated with the Kansas Commission on Veterans Affairs in Topeka, Kansas, is the paper's author. In this study, vaccinated veterans who did not serve in the Persian Gulf War showed significantly more Gulf War Illness symptoms compared to unvaccinated veterans who did not serve in the Persian Gulf War.[2] Vaccinated veterans had 1.94 times greater odds of experiencing joint pain (95% CI of 1.02 to 3.70), 3.02 times greater odds of experiencing short-term memory issues (95% CI of 1.28 to 7.11), 4.48 times greater odds of accessing words (95% CI of 1.61 to 12.48), and 3.53 times greater odds

Prevalence and Patterns of Gulf War Illness in Kansas Veterans: Association of Symptoms with Characteristics of Person, Place, and Time of Military Service

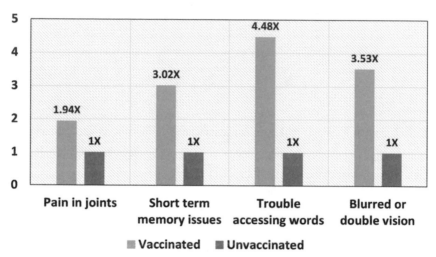

Figure 6.1—Odds ratios for Gulf War Illness symptoms in vaccinated non-Persian Gulf War veterans versus unvaccinated non-Persian Gulf War veterans (Steele 2000).

of experiencing blurred vision (95% CI of 1.13 to 11.03) than unvaccinated veterans. Veterans were considered "vaccinated" if they received any vaccines from the military between August 1990 and July 1991 and "unvaccinated" if they received no vaccines from the military in the same period.

Figure 6.2 shows results from the paper "Health of UK Servicemen Who Served in the Persian Gulf War," published in the journal *Lancet* in 1999.[3] Dr. Catherine Unwin, affiliated with the Gulf War Illness Research Unit at Guy's, King's, and St. Thomas's Medical School in London, United Kingdom, is the lead author of the paper. UK

Health of UK Servicemen Who Served in the Persian Gulf War

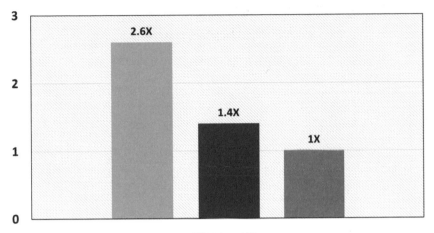

Odds Ratios for Gulf War Illness Versus Number of Vaccines Received

Figure 6.2—Odds ratios for Gulf War Illness versus the number of vaccines received by Bosnia War and Persian Gulf War servicemen from the United Kingdom (Unwin et al. 1999).

servicemen who received multiple vaccinations showed significantly more Gulf War illness symptoms than unvaccinated UK servicemen, with those who received more than seven vaccines having 2.6 times greater odds of experiencing Gulf War Illness symptoms (p-value <0.0001 and a 95% CI of 2.2 to 3.1), and those who received between three and six vaccines had 1.4 times greater odds of Gulf War Illness symptoms (p-value < 0.0001 and a 95% CI of 1.2 to 1.6) than unvaccinated servicemen.[4] The study authors stated, "Vaccination against biological warfare and multiple routine vaccinations were associated with the CDC multisymptom syndrome in the Gulf War cohort."[5] The vaccination status of these servicemen was based on vaccines received within two months before and during each conflict.

Role of Vaccinations as Risk Factors for Ill Health in Veterans
of the Gulf War: Cross-Sectional Study.

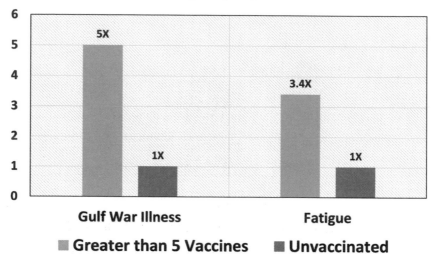

Odds Ratios for Gulf War Illness and Fatigue from Multiple Vaccines Received During Deployment

Figure 6.3—Odds ratios for Gulf War Illness and fatigue in deployed
servicemen who received multiple vaccines versus deployed, unvaccinated
servicemen (Hotopf et al. 2000).

Figure 6.3 shows results from the paper "Role of Vaccinations
as Risk Factors for Ill Health in Veterans of the Gulf War: Cross-
Sectional Study," published in *BMJ* in 2000.[6] The lead author is Dr.
Matthew Hotopf, affiliated with the Gulf War Illness Research Unit
at Guy's, King's, and St. Thomas's School of Medicine, King's College
London in London, United Kingdom. Veterans receiving multiple
vaccines during deployment were diagnosed with multisymptom Gulf
War Illness (p-value < 0.0001 and a 95% CI of 2.5 to 9.8) and fatigue
(p-value < 0.0001 and a 95% CI of 1.9 to 6.2) at a far greater frequency
than unvaccinated veterans who also served in the Gulf War.[7]

Figure 6.4 shows results from the paper "Symptoms and Medical Conditions in Australian Veterans of the 1991 Gulf War: Relation to Immunisations and Other Gulf War Exposures," published in the journal *Occupational and Environmental Medicine* in 2004.[8] Dr. H. L. Kelsall, affiliated with the Department of Epidemiology and Preventative Medicine at Monash University–Central and Eastern Clinical School in Melbourne, Australia, is the lead author. Australian veterans receiving ten or more vaccinations while in military service showed a statistically significantly increased number

Symptoms and Medical Conditions in Australian Veterans of the 1991 Gulf War: Relation to Immunisations and Other Gulf War Exposures

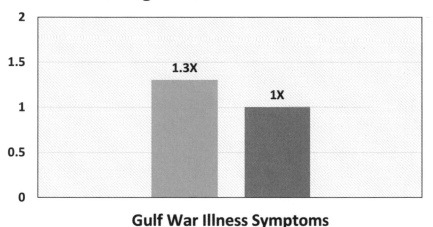

Figure 6.4—Odds ratio for Gulf War Illness Symptoms in Gulf War veterans receiving ten or more vaccinations compared to unvaccinated Gulf War veterans (Kelsall et al. 2004).

of Gulf War Illness symptoms compared to unvaccinated veterans (p-value < 0.001 and a 95% CI of 1.2 to 1.4).[9] This analysis considered the total number of symptoms reported but did not examine the severity of symptoms.

Summary

Table 6.1 shows results for the four publications highlighted in this chapter.[10, 11, 12, 13] Steele[14] and Hotopf et al.[15] included individual Gulf War Illness symptoms within their studies. However, Unwin et al.[16] focused on Gulf War Illness as a syndrome of multiple symptoms. All studies focused on the number of vaccines received either before or during deployment during military service.[17, 18, 19, 20]

	Steele 2000	Unwin et al. 1999	Hotopf et al. 2000	Kelsall et al. 2004
Gulf War Illness Symptoms		✓	✓	✓
Pain in Joints	✓			
Short term memory issues	✓			
Trouble Accessing Words	✓			
Blurred or Double Vision	✓			
Fatigue			✓	

Table 6.1—Summary of results for vaccinated versus unvaccinated veterans. Significantly higher odds ratios, relative risks, or incidences are denoted by a ✓.

CHAPTER 7

Influenza (Flu) Vaccines

In the United States, the CDC recommends the annual influenza (flu) vaccine for every child six months of age and older and every adult.[1] This recommendation includes pregnant women in any trimester of pregnancy.[2] The flu shot is available as either the trivalent inactive virus (TIV) vaccine or a live attenuated influenza virus (LAIV) vaccine. The LAIV is contraindicated for numerous conditions, including pregnancy, asthma, and immunosuppression. Some manufacturers distribute TIV vaccines in multidose vials, which contain 25 micrograms of mercury per dose in the form of thimerosal.[3] Infant formulations of the same vaccines contain 12.5 micrograms of mercury in a two-shot series for a total inoculation of 25 micrograms of mercury.[4] In addition to the seasonal flu shot, vaccine manufacturers formulated and distributed the H1N1 pandemic influenza vaccine (swine flu) between 2009 and 2011.[5] Multidose vials of the H1N1 vaccine also contained thimerosal. In this chapter, we consider both the seasonal flu shot and the H1N1 flu shot.

Figure 7.1 shows results from the paper "Risk of Narcolepsy in Children and Young People Receiving AS03 Adjuvanted Pandemic A/H1N1 2009 Influenza Vaccine: A Retrospective Analysis," published

Risk of Narcolepsy in Children and Young People Receiving ASo3 Adjuvanted Pandemic A/H1N1 2009 Influenza Vaccine: A Retrospective Analysis

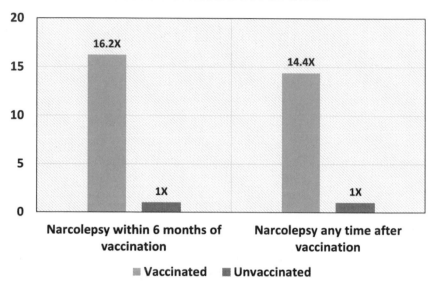

Figure 7.1—Odds ratios for narcolepsy diagnosed within six months of vaccination and at any time after vaccination with the Pandemrix H1N1 influenza vaccine (Miller et al. 2013).

in *BMJ*.[6] Dr. Elizabeth Miller, consultant epidemiologist in the Immunisation, Hepatitis and Blood Safety Department of the Health Protection Agency in London, UK, is the lead author. Miller and coauthors reported a causal association between the H1N1 vaccine and narcolepsy in children and young people in England.[7] Narcolepsy is a serious, chronic, and potentially debilitating illness characterized by a tendency to fall asleep at inappropriate times.[8] It is believed to be due to an autoimmune attack on the sleep center in the brain.[9] There is no known cure. Vaccinated individuals showed 14.4 times greater odds of a narcolepsy diagnosis any time after vaccination (95% CI of 4.3 to

48.5) compared to unvaccinated individuals.[10] If patients received the diagnosis within six months of vaccination, the odds ratio increased to 16.2 (95% CI of 3.1 to 84.5).[11] Both results were highly statistically significant.

Increased Childhood Incidence of Narcolepsy in Western Sweden after H1N1 Influenza Vaccination

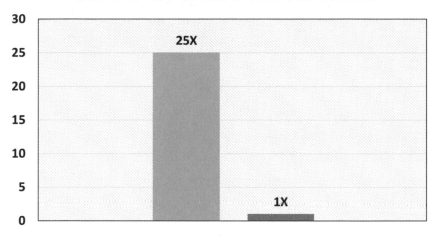

Rate of Narcolepsy in Sweden Before and After the Use of the Swine Flu Vaccine

Narcolepsy Rate

■ After Vaccine Use   ■ Before Vaccine Use

Figure 7.2—Incidence rate of narcolepsy in Sweden before and after the introduction of the Pandemrix swine flu vaccine (Szakacs et al. 2013).

Figure 7.2 shows results from the paper "Increased Childhood Incidence of Narcolepsy in Western Sweden after H1N1 Influenza Vaccination," published in the journal *Neurology*.[12] The lead author is Dr. Attila Szakacs from the Department of Pediatrics at the University of Gothenburg in Gothenburg, Sweden. Before mass vaccination, children had an incidence of narcolepsy of 0.26 out of 100,000 each year.[13] After mass vaccination, children had an incidence of narcolepsy that increased to 6.6 out of 100,000 yearly (95% CI of 3.4 to 8.1.).[14] The difference in incidence values before and after vaccination was highly statistically significant, with a p-value of less than 0.0001.

Narcolepsy Incidence and Clinical Picture of Childhood Narcolepsy Following the 2009 H1N1 Pandemic Vaccination Campaign in Finland

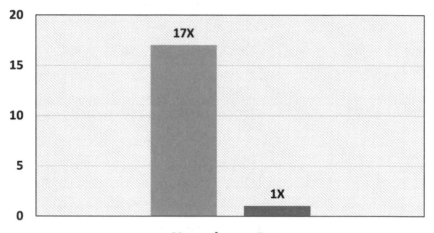

Rate of Narcolepsy in Finland Before and After the Use of the Swine Flu Vaccine

Figure 7.3—Incidence rate of narcolepsy in Finland before and after the introduction of the Pandemrix swine flu vaccine (Partinen et al. 2012).

Figure 7.3 shows results from the paper "Narcolepsy Incidence and Clinical Picture of Childhood Narcolepsy Following the 2009 H1N1 Pandemic Vaccination Campaign in Finland," published in the journal *PLoS One* in 2012.[15] Dr. Markku Partinen, from the Helsinki Sleep Clinic, Finnish Narcolepsy Research Centre, and Vitalmed Research Centre in Helsinki, Finland, is the lead author. When comparing the period before the H1N1 influenza vaccine campaign to the period after, Partinen observed a seventeenfold increase in narcolepsy in children among all sleep clinics in Finland. Before the H1N1 vaccine campaign, children had a rate of narcolepsy of only 0.31 out of 100,000 each year, and after the excessive promotion of the vaccine, children had a rate of narcolepsy that rose to 5.3 out of 100,000 each year.[16]

Neurological and Autoimmune Disorders after Vaccination against Pandemic Influenza A (H1N1) with Monovalent Adjuvanted Vaccine: Population-Based Cohort Study in Stockholm, Sweden

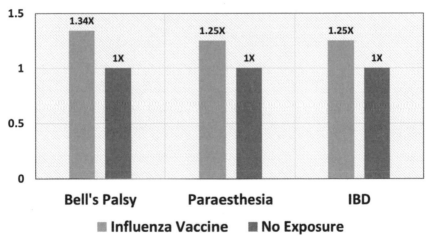

Figure 7.4—Hazard ratios for Bell's Palsy, paraesthesia, and inflammatory bowel disease in individuals receiving the H1N1 influenza vaccine versus unvaccinated individuals (Bardage et al. 2011).

Figure 7.4 shows results from the paper "Neurological and Autoimmune Disorders after Vaccination against Pandemic Influenza A (H1N1) with Monovalent Adjuvanted Vaccine: Population-Based Cohort Study in Stockholm, Sweden" in *BMJ* in 2011.[17] The lead author is Dr. Carola Bardage, an epidemiologist at the Medical Products Agency in Uppsala, Sweden. This is a cohort study of the entire populace of Stockholm County, Sweden, which had a population of nearly two million people and a 52.6% vaccination rate. Dr. Bardage saw elevated risks for Bell's Palsy (95% CI of 1.11 to 1.64), paraesthesia (an abnormal sensation of tingling or prickling) (95% CI of 1.10 to 1.41), and inflammatory bowel disease (95% CI of 1.04 to 1.50) among individuals vaccinated within 45 days of the start of the H1N1 vaccination campaign.[18] These were primarily high-risk individuals who qualified for early vaccination. However, any skewing of the results due to this type of patient was ameliorated by adjusting for differences in health-care-seeking behavior. Practitioners diagnosed patients based on inpatient and specialist utilization using the common healthcare registries for the Stockholm County Council.

Figure 7.5 shows results from the paper "Risk of Guillain-Barré Syndrome after Seasonal Influenza Vaccination and Influenza Health-Care Encounters: A Self-Controlled Study," published in the journal *Lancet Infectious Disease* in 2013.[19] Dr. Jeff Kwong, affiliated with the Institute for Clinical Evaluative Sciences in Toronto, Canada, is the lead author. Guillain-Barré syndrome (GBS) is a serious disorder in which your body's immune system attacks your nerves, leading to paralysis.[20] Recovery from GBS may take several years, and some cases are fatal.

Based on healthcare data recorded from 1993 to 2011 in Ontario, Canada, Kwong determined that the risk of GBS was 52% higher within six weeks of vaccination than it was in the control time frame of nine to forty-two weeks prior to vaccination, with a relative incidence of 1.52 and a 95% confidence interval of 1.17 to 1.99.[21] In addition, the risk of GBS within six weeks of influenza infection was much greater than after vaccination, with a relative incidence of 15.81 and a 95% confidence interval of 10.28 to 24.32.[22] However, only a small

Risk of Guillain-Barré Syndrome after Seasonal Influenza Vaccination and Influenza Health-Care Encounters: A Self-Controlled Study

Relative Risk of Guillain-Barré Syndrome Following the Seasonal Flu Shot

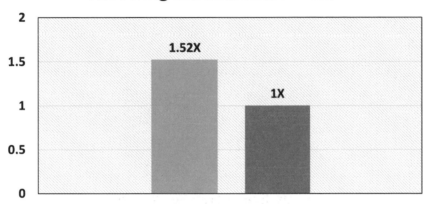

Figure 7.5—Relative risk of Guillain-Barré syndrome within six weeks of seasonal influenza vaccination compared to a control period of 9 to 42 weeks after vaccination (Kwong et al. 2013).

percentage of the population contracts the flu in any given year, while the entire population is encouraged to receive the vaccine, which has marginal effectiveness. This means that seasonal influenza vaccination could likely increase the overall rate of GBS.

Figure 7.6 shows the results of the paper "Guillain-Barré Syndrome after Influenza Vaccination in Adults: A Population-Based Study," published in *JAMA Internal Medicine* in 2006.[23] Dr. David Juurlink, affiliated with the Institute of Clinical Evaluative Sciences in Toronto, Canada, is the lead author. Like the previous study,[24] Juurlink reported an increased incidence of GBS following seasonal influenza

Guillain-Barré Syndrome after Influenza Vaccination in Adults:
A Population-Based Study

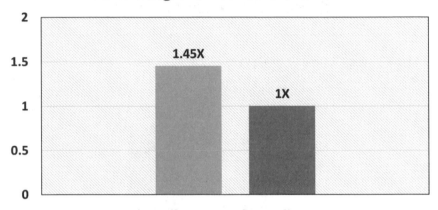

Figure 7.6—Relative risk of Guillain-Barré syndrome between 2 and 7 weeks after seasonal influenza vaccination compared to a control period of 20 to 43 weeks after vaccination (Juurlink et al. 2006).

vaccination.[25] In this study, researchers investigated 1,601 hospital admissions due to GBS in Ontario, Canada, and the relative incidence of the illness following the flu shot was 1.45, with a p-value of 0.02 and a 95% confidence interval of 1.05 to 1.99.[26]

Figure 7.7 shows results from the paper "The Guillain-Barré Syndrome and the 1992–1993 and 1993–1994 Influenza Vaccines," published in *The New England Journal of Medicine* in 1998.[27] The lead author is Dr. Tamar Lasky, from the Department of Epidemiology and Preventative Medicine at the School of Medicine, University of Maryland in Baltimore. In the US, between 1992 and 1994, Dr. Lasky

The Guillain-Barré Syndrome and the 1992–1993 and 1993–1994 Influenza Vaccines

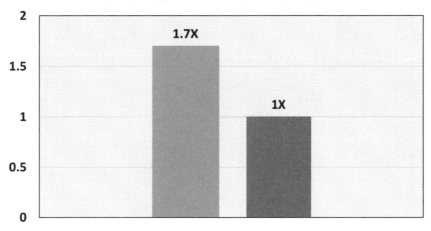

Figure 7.7—Relative risk of Guillain-Barré syndrome within six weeks after seasonal influenza vaccination compared to a control group of all others in the cohort (Lasky et al. 1998).

observed an overall relative incidence of GBS of 1.7 with a p-value of 0.04 and a 95% confidence interval of 1.0 to 2.8 following seasonal flu vaccines.[28]

Figure 7.8 shows results from the paper "Guillain-Barré Syndrome during the 2009–2010 H1N1 Influenza Vaccination Campaign: Population-Based Surveillance among 45 Million Americans," published in the *American Journal of Epidemiology* in 2012.[29] The lead author is Dr. Matthew Wise, affiliated with the Division of Health Quality Promotion at the CDC in Atlanta, Georgia. When investigating the relationship between the H1N1 vaccine distributed in the US

Guillain-Barré Syndrome during the 2009–2010 H1N1
Influenza Vaccination Campaign: Population-Based
Surveillance among 45 Million Americans

Rate Ratio of Guillain-Barré Syndrome Following the H1N1 Flu Shot

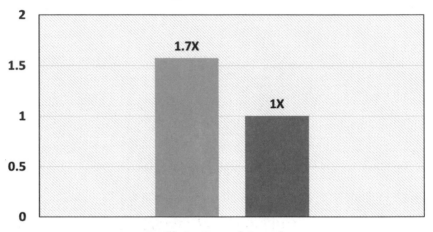

Guillain Barré Syndrome

■ **Vaccinated**   ■ **Unvaccinated**

Figure 7.8—Relative risk of Guillain-Barré syndrome after H1N1 influenza
vaccination compared to unvaccinated patients (Wise et al. 2012).

from 2009 to 2010, Dr. Wise discovered a 57% increase in cases of GBS
among vaccinated individuals compared to unvaccinated individuals
(95% CI of 1.02 to 2.21).[30]

Figure 7.9 shows results from the paper "The Risk of Guillain-Barré
Syndrome Associated with Influenza A (H1N1) 2009 Monovalent
Vaccine and 2009–2010 Seasonal Influenza Vaccines: Results from Self-
Controlled Analyses," published in the journal *Pharmacoepidemiology
and Drug Safety* in 2012. Dr. Jerome Tokars, from the Division of
Health Quality Promotion at the CDC in Atlanta, Georgia, is the lead
author. Following the distribution of the H1N1 vaccine, Dr. Tokars
detected an increased risk of GBS of 3.0 with a 95% confidence interval

The Risk of Guillain-Barré Syndrome Associated with Influenza A (H1N1) 2009 Monovalent Vaccine and 2009–2010 Seasonal Influenza Vaccines: Results from Self-Controlled Analyses

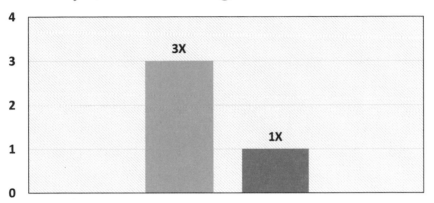

Relative Risk of Guillain-Barré Syndrome Following the H1N1 Flu Shot

Guillain-Barré Syndrome
▇ Within 42 Days of Vaccination
▇ Between 43 and 84 Days from Vaccination

Figure 7.9—Relative risk of Guillain-Barré syndrome diagnosed within 42 days for H1N1 influenza vaccination compared to between 43 and 84 days from vaccination (Tokars et al. 2012).

of 1.4 to 6.4 in a self-controlled study comparing diagnoses within 42 days of vaccination to diagnoses made between 43 and 84 days from vaccination.[31]

Figure 7.10 shows results from the paper "Association between Guillain-Barré Syndrome and Influenza A (H1N1) 2009 Monovalent Inactivated Vaccines in the USA: A Meta-Analysis," published in the journal *Lancet* in 2013.[32] The lead author is Dr. Daniel Salmon, affiliated with the National Vaccine Program Office of the US Department of Health and Human Services in Washington, DC. Using a self-controlled analysis where the control group consisted of vaccinated

Association between Guillain-Barré Syndrome and Influenza
A (H1N1) 2009 Monovalent Inactivated Vaccines in the USA:
A Meta-Analysis

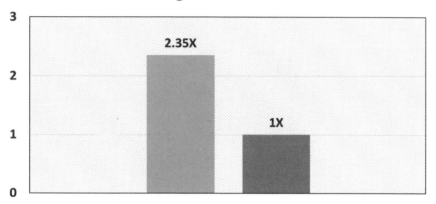

Relative Risk of Guillain-Barré Syndrome Following the H1N1 Flu Shot

Guillain-Barré Syndrome

■ Within 42 Days of Vaccination

■ After 43 Days from Vaccination

Figure 7.10—Relative risk of Guillain-Barré syndrome diagnosed within 42 days
for H1N1 influenza vaccination compared to after 43 days from vaccination
(Salmon et al. 2013).

individuals followed starting 43 days from vaccination, Dr. Salmon
also noted an associated increased risk of GBS of 2.35, with a p-value
of 0.0003 and a 95% confidence interval of 1.42 to 4.01, within 42 days
of the H1N1 vaccine.[33]

Figure 7.11 shows results from the paper "Assessment of Temporally
Related Acute Respiratory Illness following Influenza Vaccination,"
published in *Vaccine* in 2018.[34] Dr. Sharon Rikin, with the Department
of Medicine at Columbia University in New York, New York, is the
lead author. Vaccinated children four years old and younger showed a
4.8 times greater hazard of non-influenza acute respiratory infection

Assessment of Temporally Related Acute Respiratory Illness following Influenza Vaccination

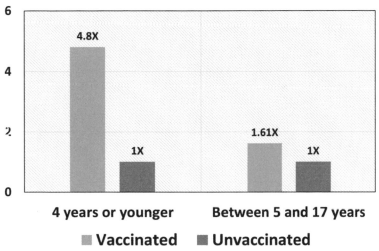

Figure 7.11—Hazard ratios for acute respiratory infection for children vaccinated for seasonal influenza versus unvaccinated children (Rikin et al. 2018).

compared to unvaccinated peers within a 14-day window following vaccination.[35] This result was statistically significant with a 95% confidence interval of 2.88 to 7.99. Vaccinated children between 5 and 17 years old showed a 1.61 times greater hazard of non-influenza acute respiratory infection compared to unvaccinated peers.[36] This result was marginally significant, with a 95% confidence interval of 0.98 to 2.66.[37]

Figure 7.12 shows results from the paper "Influenza Vaccination and Respiratory Virus Interference among Department of Defense Personnel during the 2017–2018 Influenza Season," published in *Vaccine* in 2020.[38] Dr. Greg Wolff, affiliated with the Armed Forces Health Surveillance Branch, Air Force Satellite at Wright-Patterson Air Force Base in Ohio, is the author. Virus interference occurs when

Influenza Vaccination and Respiratory Virus Interference among Department of Defense Personnel during the 2017–2018 Influenza Season

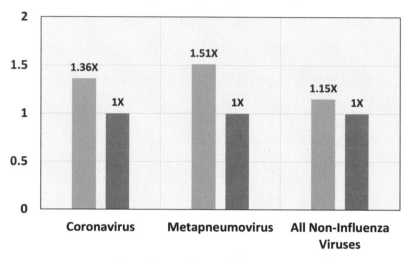

Figure 7.12—Odds ratios for coronavirus, metapneumovirus, and all non-influenza viruses when comparing seasonal influenza vaccinated to unvaccinated service members (Wolff 2020).

vaccinated individuals may be at greater risk for other viruses because they do not receive the nonspecific immunity associated with natural infection. In this analysis, vaccinated servicemen had 36% greater odds of contracting a coronavirus (p-value <0.01 and a 95% CI of 1.14 to 1.63), 51% greater odds of contracting a metapneumovirus (a virus isolated in 2001 that causes lower and upper respiratory infections) (p-value < 0.01 and a 95% CI of 1.20 to 1.90), and 15% greater odds of contracting any non-influenza virus associated with a respiratory infection (p-value < 0.01 and a 95% CI of 1.05 to 1.27).[39] All these relationships were statistically significant.

Increased Risk of Noninfluenza Respiratory Virus Infections
Associated with Receipt of Inactivated Influenza Vaccine

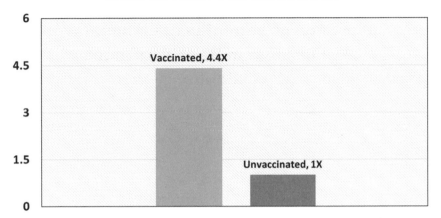

Vaccinated vs. Unvaccinated Relative Risk of Non-Flu Infections

Figure 7.13—Relative risk of non-influenza infections among vaccinated versus unvaccinated children (Cowling et al. 2012).

Figure 7.13 shows results from the study "Increased Risk of Noninfluenza Respiratory Virus Infections Associated with Receipt of Inactivated Influenza Vaccine," published in *Clinical Infectious Diseases* in 2012.[40] The lead author is Dr. Benjamin Cowling, affiliated with the School of Public Health, Li Ka Shing Faculty of Medicine, University of Hong Kong in China. In this randomized, prospective study, 115 children between the ages of six and 15 received either the trivalent inactivated vaccine or a placebo. Researchers monitored these children for nine months after the injection. The relative risk of non-influenza respiratory infection was 4.40 when comparing vaccinated to unvaccinated children, with a 95% confidence interval of 1.31 to 14.8.[41] Vaccinated and placebo groups showed no statistically significant difference in the incidence of influenza, with a relative risk of 0.66 and a 95% confidence interval of 0.13 to 3.27.[42] This may be due to the low number of influenza cases ascertained overall.

Epidemiology of Respiratory Viral Infections in Children Enrolled in a Study of Influenza Vaccine Effectiveness

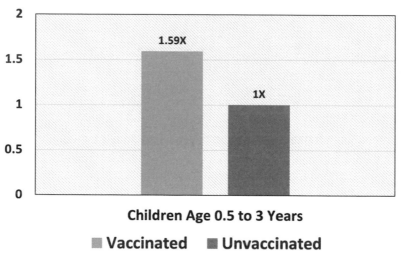

Figure 7.14—Rate ratio of non-influenza influenza-like infections among H1N1 vaccinated versus unvaccinated children (Dierig et al. 2014).

Figure 7.14 shows results from the paper "Epidemiology of Respiratory Viral Infections in Children Enrolled in a Study of Influenza Vaccine Effectiveness," published in the journal *Influenza and Other Respiratory Viruses* in 2014.[43] Dr. Alexa Dierig, affiliated with the National Centre for Immunisation Research and Surveillance at the Children's Hospital of Westmead in Westmead, Australia, is the lead author. This study examined H1N1 vaccinated and unvaccinated children during the thirteen-week influenza season in 2010. The analyzed cohort was comprised of 381 children, with 238 unvaccinated and 143 vaccinated, who contracted 124 influenza-like illnesses over the thirteen-week period.[44] Influenza-like illnesses that practitioners diagnosed included H1N1 influenza, NL63 coronavirus, and, most frequently, adenoviruses and rhinoviruses. Accordingly, the authors discovered that vaccinated children were 1.59 times more likely to have a non-influenza influenza-like illness than unvaccinated children, with a p-value of 0.001.[45]

Effectiveness of Trivalent Inactivated Influenza Vaccine in Influenza-Related Hospitalization in Children: A Case-Control Study

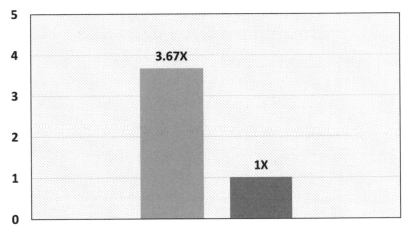

Figure 7.15—Relative number of influenza-related hospitalizations in children receiving the influenza vaccine compared to unvaccinated children (Joshi et al. 2012).

Figure 7.15 shows results from the paper "Effectiveness of Trivalent Inactivated Influenza Vaccine in Influenza-Related Hospitalization in Children: A Case-Control Study," published in the journal *Allergy and Asthma Proceedings* in 2012.[46] The lead author is Dr. Avni Joshi, affiliated with the Mayo Clinic School of Medicine in Rochester, Minnesota. In a study of pediatric patients, Dr. Joshi monitored patients over a seven-year period for hospital admission, emergency room visits, and severity of asthma diagnosed in a hospital setting. The researchers found that the trivalent inactivated flu vaccine ironically increased the rate of influenza hospitalizations in children by 3.67 times, with a statistically significant 95% confidence interval of 1.6 to 8.4.[47] There was also a significant association between hospitalization in asthmatic subjects and the trivalent inactivated flu vaccine (p=0.001).[48]

Inflammation-Related Effects of Adjuvanted Influenza A on Platelet Activation and Cardiac Autonomic Function

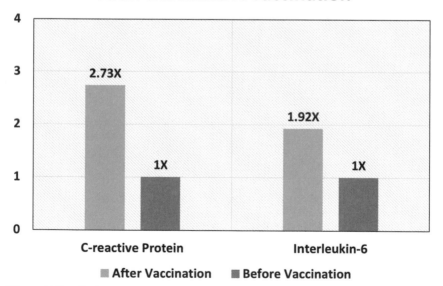

Figure 7.16—Relative level of inflammatory markers before and after influenza A vaccination (Lanza et al. 2011).

Figure 7.16 shows results from the paper "Inflammation-Related Effects of Adjuvanted Influenza A on Platelet Activation and Cardiac Autonomic Function," published in the *Journal of Internal Medicine in 2011.*[49] The lead author is Dr. Gaetano A. Lanza, affiliated with the Instituto di Cardiologia at the Università Cattolica del Sacro Cuore in Rome, Italy. The investigators observed 28 individuals with type 2 diabetes inoculated with the adjuvanted influenza A vaccine. Before and after vaccination, researchers measured C-reactive protein (CRP), interleukin 6, and monocyte-platelet aggregates. After vaccination, the patients' CRP level elevated from 2.6 to 7.1 milligrams per liter with a p-value less than 0.0001, interleukin 6 increased from 0.82 to 1.53 picograms per milliliter with a p-value less than 0.0001, and mono-cyte-platelet aggregation increased from 28.5% to 30.5%.[50] These results demonstrate a direct connection between the inflammatory stimulus of vaccination and platelet activation, as well as a direct connection between inflammatory stimulus and cardiac autonomic activity. Correlations found between changes in CRP level and heart rate variability suggest a pathophysiological link between the inflammatory and cardiac auto-nomic responses to vaccine administration. Pathophysiology refers to the disordered physiological processes associated, in this case, with cardiac disease. The increased platelet activation observed after vacci-nation could transiently increase the likelihood of thrombosis (local-ized clotting of the blood) in high-risk patients. Thus, vaccine-induced changes in platelet activity and autonomic nervous activity may tempo-rarily increase the risk of cardiovascular events in vaccinated patients. Interestingly, there are 17,922 reported events of "cardiomyopathy" asso-ciated with influenza vaccines in the VAERS database.[51]

Summary

Three research papers showed significant relationships between the seasonal influenza vaccine and GBS.[52, 53, 54] The seasonal flu shot also

	Kwong et al. 2013	Juurlink et al. 2006	Lasky et al. 1998	Rikin et al. 2018	Wolff 2020	Cowling et al. 2012	Joshi et al. 2012	Lanza et al. 2011
Guillain-Barré Syndrome	✓	✓	✓					
Acute Respiratory Infections				✓				
Coronavirus					✓			
Metapneumovirus					✓			
All Non-Influenza Viruses					✓	✓		
Hospitalizations Due to Influenza							✓	
Inflammation Markers								✓

Table 7.1 - Summary of results comparing health outcomes of individuals exposed to the seasonal influenza vaccine. Significantly higher odds ratios, relative risks, hazard ratios, or incidences are denoted by a ✓.

	Miller et al. 2013	Szakacs et al. 2013	Partinen et al. 2012	Bardage et al. 2011	Wise et al. 2012	Tokars et al. 2012	Salmon et al. 2013	Dierig et al. 2014
Narcolepsy	✓	✓	✓					
Bell's Palsy				✓				
Paraesthesia				✓				
IBD				✓				
Guillain-Barré Syndrome					✓	✓	✓	
Influenza-Like Infections								✓

Table 7.2 - Summary of results in comparing health outcomes of individuals exposed to the H1N1 pandemic influenza vaccine. Significantly higher odds ratios, relative risks, hazard ratios, or incidences are denoted by a ✓.

correlated with an uptick in non-influenza respiratory viruses in two studies.[55, 56] Investigators highlighted acute respiratory infections,[57] coronavirus,[58] metapneumovirus,[59] hospitalization due to influenza,[60] and inflammation markers[61] in each of the research studies considered.

Three research papers showed a significant relationship between the H1N1 influenza vaccine and narcolepsy [62, 63, 64] and GBS.[65, 66, 67] The three papers focused on narcolepsy as an adverse effect of the H1N1 vaccine and considered cohorts in the UK, Finland, and Sweden when the Pandemrix® vaccine was distributed.[68, 69, 70] In addition, one paper each considered outcomes that included Bell's palsy,[71] paraesthesia,[72] inflammatory bowel disease,[73] and influenza-like infections.[74]

CHAPTER 8

DTP Vaccines

Vaccine manufacturers withdrew the diphtheria-tetanus-whole cell pertussis (DTP) vaccine from the US market in the 1980s and 1990s due to the high frequency and extreme severity of vaccine adverse events.[1] The diphtheria-tetanus-acellular pertussis (DTaP) vaccine is given instead. Although the DTP is no longer used in the US, vaccine manufacturers distribute it throughout other parts of the world including Africa, Asia, and Central and South America. It is often combined with hepatitis B and *Haemophilus influenzae* type b (Hib) in a pentavalent vaccine.[2] Although we could find no vaccinated versus unvaccinated studies that investigated the DTaP vaccine, we located several for the DTP.

These articles focused on nonspecific effects (NSEs) of vaccines also referred to as "off-target" effects. NSEs are effects that are outside the protection of vaccines against the targeted pathogens. They differ from side effects, which refer to undesirable localized reactions at the injection site (such as tenderness, swelling, pain, and bruising) or systemic reactions (such as fever, rash, joint, and muscle pain) that typically resolve within days or weeks.[3] NSEs could be theoretically beneficial and increase the ability of other vaccines to protect against targeted pathogens or even nontargeted pathogens. In other circumstances, however, NSEs may be harmful, enhancing susceptibility to illness or even death due to causes other than the targeted infections.

According to the World Health Organization (WHO) and various researchers, vaccination sequence and dosage recommendations are critical considerations in ensuring optimized vaccine protection. Unfortunately, scientists have conducted few studies to determine even the baseline mortality rates in the vaccinated and unvaccinated populations to establish whether total mortality rates increase or decrease due to vaccine administration. Apart from the specific-pathogen protective effect of vaccines, inherent bias has led to the belief that vaccinated children have better survival rates than unvaccinated children.[4] However, evidence from isolated studies, particularly on Bacillus Calmette-Guerin (BCG, i.e., tuberculosis) and DTP-containing vaccines within developing countries, showed NSEs' role in increased mortality rates among the vaccinated.[5] Based on the findings, the WHO ordered a review of NSEs associated with BCG, DTP, and live measles vaccine (MV) in 2013.[6] The findings confirmed a link between DTP-containing vaccination sequences and NSEs in high-mortality regions.[7]

This chapter features six brilliant studies by Dr. Peter Aaby and his research associates. Dr. Aaby was one of the first scientists to complete research on the NSEs of vaccines, publishing research papers on the topic as early as 2000. He focused his study on children in the rural areas of Guinea-Bissau, Africa, and looked at the relationship between the DTP vaccine (among others, given in combination) and infant mortality following vaccination. Dr. Aaby found that contrary to the intent of the vaccination programs in underserved countries, such as Guinea-Bissau, infant mortality was higher, specifically in those children inoculated with DTP.

Other studies in this chapter focus on sudden infant death

syndrome, allergies, asthma, and eczema associated with the DTP vaccine.

Figure 8.1 presents results from the paper "The Introduction of Diphtheria-Tetanus-Pertussis and Oral Polio Vaccine among Young Infants in an Urban African Community: A Natural Experiment," published in the journal *EBioMedicine* in 2017.[8] The lead author, Dr. Soren Mogensen, was affiliated with the Bandim Health Project in Guinea-Bissau, Africa. The corresponding author, Dr. Peter Aaby, is a professor in the Department of Clinical Research at the University of Southern Denmark in Odense, Denmark. Researchers followed unvaccinated children and children vaccinated with the DTP vaccine

The Introduction of Diphtheria-Tetanus-Pertussis and Oral Polio Vaccine among Young Infants in an Urban African Community: A Natural Experiment

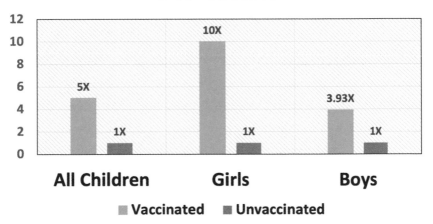

Figure 8.1—Hazard ratio for mortality in children vaccinated with the DTP vaccine compared to unvaccinated children in Guinea-Bissau, Africa (Mogensen et al. 2017).

Early Diphtheria-Tetanus-Pertussis Vaccination Associated with
Higher Female Mortality and No Difference in Male Mortality
in a Cohort of Low Birthweight Children: An Observational
Study within a Randomized Trial

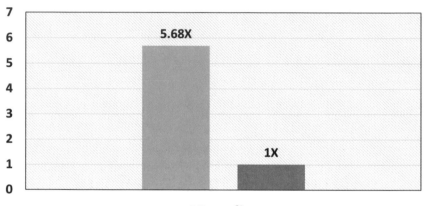

Figure 8.2—Infant mortality in girls in Guinea-Bissau who received one early
DTP vaccine compared to no DTP vaccines (Aaby et al. 2012).

in Guinea-Bissau between three and five months of age. Vaccinated
children showed five times higher mortality than unvaccinated children (95% CI of 1.53 to 16.3), with the most dramatic results observed
in girls (95% CI of 0.81 to 123.0).[9]

Figure 8.2 shows results from the paper "Early Diphtheria-Tetanus-
Pertussis Vaccination Associated with Higher Female Mortality and
No Difference in Male Mortality in a Cohort of Low Birthweight
Children: An Observational Study within a Randomized Trial,"
published in the journal *Archives of Disease in Children* in 2012.[10] The
lead author is Dr. Peter Aaby. Using the data from Guinea-Bissau, the

The Introduction of Diphtheria-Tetanus-Pertussis Vaccine and Child Mortality in Rural Guinea-Bissau: An Observational

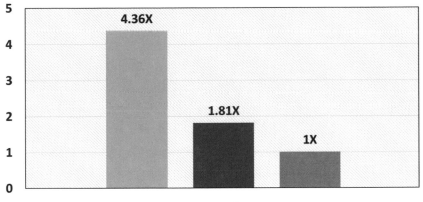

Infant Mortality in Children Receiving the First or Second/Third Dose of the DTP Versus Unvaccinated Children

Mortality

■ Second/Third DTP Dose ■ First DTP Dose ■ Unvaccinated

Figure 8.3—Infant mortality in children in rural Guinea-Bissau who received doses of the DTP vaccine versus unvaccinated children (Aaby et al. 2004).

researchers found that girls who received the DTP at two months were 5.68 times more likely to die between two-month and six-month practitioner visits than girls who had not received the DTP vaccine (95% CI of 1.83 to 17.7).[11] For boys and girls combined, the mortality rate was 2.62 times higher in vaccinated children (95% CI of 1.34 to 5.09).[12]

Figure 8.3 shows results from the paper "The Introduction of Diphtheria-Tetanus-Pertussis Vaccine and Child Mortality in Rural Guinea-Bissau: An Observational Study," published in the *International Journal of Epidemiology* in 2004.[13] The lead author is Dr. Peter Aaby. Children who received two or three doses of the DTP vaccine between two and eight months of age showed the highest mortality rate, 4.36 times that of the unvaccinated children (95% CI of 1.28 to 14.9), followed by those who received only one dose of the DTP vaccine with a mortality rate of 1.81 times the rate for unvaccinated children (95% CI of 0.95 to 3.45).[14] Children receiving the BCG (tuberculosis) vaccine exhibited slightly lower mortality, although differences between these children and unvaccinated children were not statistically significant.[15]

Figure 8.4 shows results from the paper "Is Diphtheria-Tetanus-Pertussis (DTP) Associated with Increased Female Mortality? A Meta-Analysis Testing the Hypothesis of Sex-Differential Non-Specific Effects of DTP Vaccine," published in the journal *Transactions of the Royal Society of Tropical Medicine and Hygiene* in 2016.[16] The lead author is, again, Dr. Peter Aaby. The authors investigated seven separate studies

Is Diphtheria-Tetanus-Pertussis (DTP) Associated with Increased Female Mortality? A Meta-Analysis Testing the Hypothesis of Sex-Differential Non-Specific Effects of DTP Vaccine

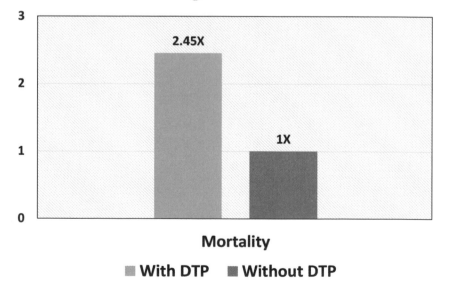

Figure 8.4—Mortality ratio in DTP-vaccinated girls who previously received the BCG (tuberculosis) vaccine (Aaby et al. 2016).

of BCG-vaccinated children and found that girls who also received the DTP vaccine had a large, statistically significant increase in mortality (95% CI of 1.48 to 4.06), with no increase in mortality in boys.[17] This outcome countered a report commissioned by WHO that posited there was no convincing evidence of such a relationship.[18]

Figure 8.5 shows results from the paper "Routine Vaccinations and Child Survival: Follow-Up Study in Guinea-Bissau, West Africa," published in *BMJ* in 2000.[19] The lead author is Dr. Ines Kristensen from the Bandim Health Project in Guinea-Bissau, Africa. The corresponding author is Dr. Peter Aaby. Children who received a single DTP or polio vaccine were 1.84 times more likely to die compared

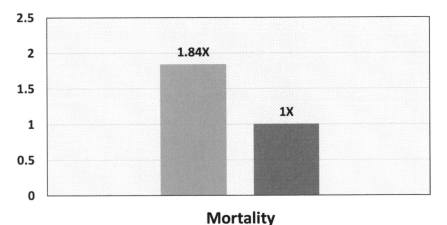

**Routine Vaccinations and Child Survival:
Follow-Up Study in Guinea-Bissau, West Africa**

Infant Mortality in Children Receiving One DTP or Polio Vaccine Versus No DTP or Polio Vaccines

Figure 8.5—Infant mortality ratio for children receiving one DTP or polio vaccination versus children receiving no DTP or polio vaccinations (Kristensen et al. 2000).

to children who received neither of those vaccines (95% CI of 1.10 to 3.10).[20] However, when considering any type of vaccine received in infancy, there was no statistically significant difference in infant mortality between vaccinated and unvaccinated children.[21]

Figure 8.6 shows results from the paper "Sex-Differential and Non-Specific Effects of Routine Vaccinations in a Rural Area with Low Vaccination Coverage: An Observational Study from Senegal," published in the journal *Transactions of the Royal Society of Tropical Medicine and Hygiene* in 2015.[22] The lead author is Dr. Peter Aaby. This study considered 4,133 children born between 1996 and 1999. Children

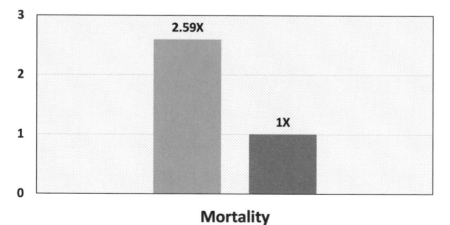

Sex-Differential and Non-Specific Effects of Routine Vaccinations in a Rural Area with Low Vaccination Coverage: An Observational Study from Senegal

Figure 8.6—Infant mortality among children vaccinated with DTP and measles virus (MV) simultaneously or DTP after MV versus the MV vaccine alone (Aaby et al. 2015).

who received the DTP and live measles vaccine simultaneously or the DTP after the measles vaccine had significantly higher mortality compared to children having the measles vaccine only as their most recent vaccination (95% CI of 1.32 to 5.07).[23] Here, Dr. Aaby and his colleagues investigated the timing of administration of measles live virus vaccine and inactivated DTP vaccine. In general, from a compilation of several publications, they found that vaccinating with live virus vaccines after inactivated vaccines led to lower mortality rates in children.

Figure 8.7 shows results from the paper "Evaluation of Non-Specific Effects of Infant Immunization on Early Infant Mortality in a Southern

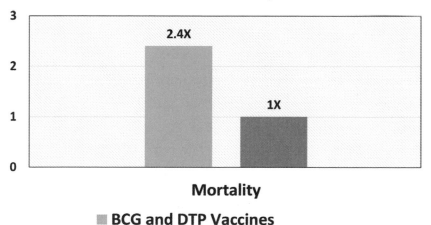

Evaluation of Non-Specific Effects of Infant Immunization on Early Infant Mortality in a Southern Indian Population

Infant Mortality in Girls Receiving Both BCG and DTP Vaccines Versus One of the Vaccines Only

Figure 8.7—Infant mortality in girls receiving both the BCG and DTP vaccines versus one of either vaccine only (Moulton et al. 2005).

Indian Population," published in the journal *Tropical Medicine and International Health* in 2005.[24] The lead author is Dr. Lawrence H. Moulton, affiliated with the Department of International Health at the Johns Hopkins Bloomberg School of Public Health in Baltimore, Maryland. In this study of 10,274 infants in Southern India, the investigators found that girls who received both the BCG and DTP vaccines had a mortality ratio of 2.4 times that of girls who received just one of either vaccine (95% CI of 1.2 to 5.0).[25]

Figure 8.8 shows results from the paper "Diphtheria-Tetanus-Pertussis Immunization and Sudden Infant Death Syndrome" published in the *American Journal of Public Health* in 1987.[26] The lead

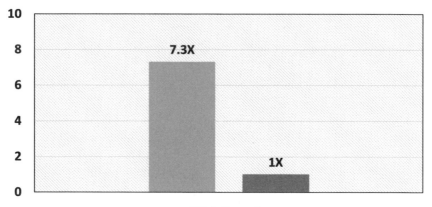

Figure 8.8—Sudden infant death syndrome (SIDS) deaths reported within three days of DTP vaccination compared to SIDS deaths reported starting thirty days after vaccination (Walker et al. 1987).

author is Dr. Alexander M. Walker, affiliated with the Boston University Medical Center in Waltham, Massachusetts, and the Harvard School of Public Health in Cambridge, Massachusetts. Researchers studied US children born between 1972 and 1983 who received the diphtheria-tetanus-whole cell pertussis vaccine. In this cohort, infants weighing more than 2,500 grams at birth experienced 7.3 times more Sudden Infant Death Syndrome (SIDS) within three days of DTP vaccination than in a period starting 30 days after DTP vaccination (95% CI of 1.7 to 31).[27]

Figure 8.9 shows results from the conference abstract "Diphtheria-Pertussis-Tetanus (DPT) Immunization: A Potential Cause of Sudden

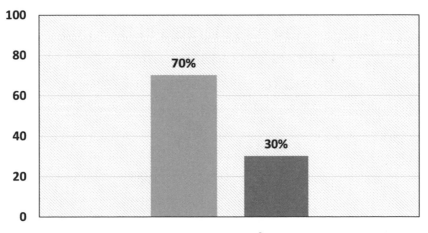

Diphtheria-Pertussis-Tetanus (DPT) Immunization: A Potential Cause of Sudden Infant Death Syndrome

Figure 8.9—SIDS deaths associated with recent DTP vaccination versus SIDS deaths without DTP vaccination (Torch 1982).

Infant Death Syndrome," presented at the American Academy of Neurology Conference in 1982.[28] The abstract's author, Dr. William C. Torch, is a pediatric neurologist in Reno, Nevada. In a study of 70 SIDS cases reported in Nevada, Dr. Torch found that 70% occurred within three weeks of receiving the DTP vaccine.[29] He also observed that SIDS cases significantly clustered within two to three weeks of DTP vaccination.[30]

Figure 8.10 shows results from the paper "Effects of Diphtheria-Tetanus-Pertussis or Tetanus Vaccination on Allergies and Allergy-Related Respiratory Symptoms among Children and Adolescents in the United States," published in the *Journal of Manipulative and Physiological Therapeutics* in 2000.[31] Dr. Eric L. Hurwitz is the lead author, affiliated with the UCLA School of Public Health in Los

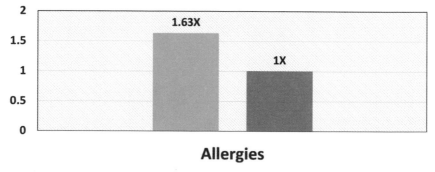

Effects of Diphtheria-Tetanus-Pertussis or Tetanus Vaccination on Allergies and Allergy-Related Respiratory Symptoms among Children and Adolescents in the United States

Odds Ratio for Allergies of DTP and Tetanus Vaccinated Children Compared to Unvaccinated Children

Figure 8.10—Odds ratio for allergies in DTP- and tetanus-vaccinated children compared to unvaccinated children (Hurwitz et al. 2000).

Angeles, California, and the Los Angeles College of Chiropractic in Whittier, California. Using data from the Third National Health and Nutrition Examination Survey on infants two months of age to adolescents sixteen years of age, the study authors investigated allergy-related respiratory symptoms over 12 months. These researchers found that DTP and tetanus-vaccinated children exhibited 63% more allergy symptoms than children who did not receive DTP or tetanus vaccinations.[32] The difference between the two groups was statistically significant, with a 95% confidence interval of 1.05 to 2.54.[33]

Figure 8.11 shows results from the paper "Delay in Diphtheria, Pertussis, Tetanus Vaccination Is Associated with a Reduced Risk of

Delay in Diphtheria, Pertussis, Tetanus Vaccination Is Associated with a Reduced Risk of Childhood Asthma

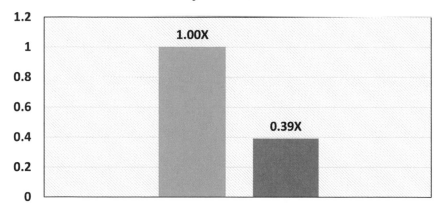

Figure 8.11—Relative risk of asthma following the recommended DTP vaccination schedule versus a delayed vaccination schedule (McDonald et al. 2008).

Childhood Asthma," published in the *Journal of Allergy and Clinical Immunology* in 2008.[34] The lead author is Kara L. McDonald, affiliated with the Faculty of Medicine at the University of Manitoba in Winnipeg, Manitoba, Canada. Dr. Anita L. Kozyrskyj, also affiliated with the Faculty of Medicine at the University of Manitoba, is the corresponding author. Among 11,531 Canadian children in this study, those who delayed the first three DTP vaccinations by more than two months showed an asthma risk of 0.39 times compared to those receiving them on time (for a total risk reduction of 61%).[35] This result was statistically significant, with a 95% confidence interval of 0.18 to 0.86.[36] In addition, children who delayed just the first DTP vaccination showed an asthma risk of 0.5 times compared to children receiving the first three vaccines on time (for a total risk reduction of 50%).[37] According to the Canadian child vaccination schedule at the time, DTP vaccines were given at two months, four months, six months, and eighteen months.

Figure 8.12 presents results from the paper "Vaccination and Allergic Disease: A Birth Cohort Study," published in the *American Journal of Public Health* in 2004.[38] The lead author is Dr. Tricia McKeever, affiliated with the University of Nottingham in the UK. Within a cohort of 29,238 UK children aged between 0 and 11 years, those children who received at least one DPPT (diphtheria, whole cell pertussis, polio, and tetanus) vaccine were 14 times as likely to be diagnosed with asthma (with a 95% CI of 7.3 to 26.9).[39] Within the same cohort, DPPT-vaccinated children were 9.4 times as likely to be diagnosed with eczema (with a 95% CI of 5.92 to 14.92).[40] The study authors claimed these results were due to differences in healthcare-seeking behavior where unvaccinated children see their practitioners less frequently. Based on their analysis of the medical records from eight large health maintenance organizations, Dr. Jason M. Glanz (of Kaiser Permanente Colorado) and his coauthors (primarily from the CDC)

Vaccination and Allergic Disease: A Birth Cohort Study

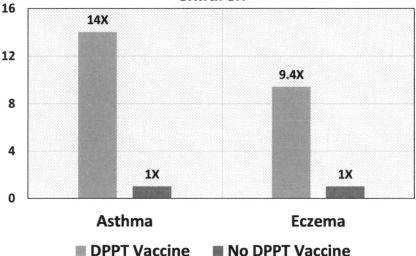

Figure 8.12: Hazard ratios for asthma and eczema diagnoses in children vaccinated with the DPPT (diphtheria-pertussis-polio-tetanus) vaccine versus unvaccinated children (McKeever et al. 2004).

reported that "under-vaccinated" children showed significantly lower outpatient provider visits.[41] However, the reported difference was a mere 10% (incidence risk ratio 0.89, 95% CI 0.89 to 0.90),[42] which is insufficient to explain the dramatic increases in asthma and eczema incidence found by Dr. McKeever.[43]

Summary

Table 8.1 summarizes the results of the twelve publications highlighted in Chapter 8. The DTP vaccine showed higher infant mortality in seven studies.[44, 45, 46, 47, 48, 49, 50] Dr. Peter Aaby coauthored six of them, based on his research in Guinea-Bissau and Senegal, Africa.[51, 52, 53, 54, 55, 56] The seventh paper was based on a cohort of children in India.[57]

	Mogensen et al. 2017	Aaby et al. 2012	Aaby et al. 2004	Aaby et al. 2016	Kristensen et al. 2000	Aaby et al. 2015	Moulton et al. 2005	Walker et al. 1987	Torch 1982	Hurwitz et al. 2000	McDonald et al. 2008	McKeever et al. 2004
Infant Mortality	✓	✓	✓	✓	✓	✓	✓					
SIDS								✓	✓			
Allergies										✓		
Asthma											✓	✓
Eczema												✓

Table 8.1 - Summary of results comparing health outcomes of children exposed to DTP (whole cell pertussis) vaccines. Significantly higher odds ratios, relative risks, or incidences are denoted by a ✓.

The DTP vaccine showed higher SIDS incidence in two publications, including a research paper by Walker in 1987[58] and an abstract presented by Torch in 1982.[59] The remaining three publications showed significant relationships between the DTP vaccination and allergies,[60] asthma,[61, 62] and eczema.[63]

CHAPTER 9

Hepatitis B Vaccines

The CDC's US childhood vaccination schedule has included the hepatitis B vaccine since the 1990s. It recommends that medical practitioners give the first dose (of a series of three shots) on the first day of life.[1] Unfortunately, there is a paucity of scientific information regarding the safety of the birth dose of the hepatitis B vaccine. However, several vaccinated versus unvaccinated studies investigate adverse events associated with hepatitis B vaccine when it is given at other stages of life.

Figure 9.1 shows the results from the paper "Hepatitis B Vaccine and Liver Problems in U.S. Children Less Than 6 Years Old," published in the journal *Epidemiology in 1999*.[2] The lead author is Dr. Monica A. Fisher, affiliated with the Department of Epidemiology at the University of Michigan in Ann Arbor. Within this study of 5,505 children participating in the 1993 National Health Information Survey, children under six years of age who received at least one dose of the hepatitis B vaccine were 2.94 times more likely to be diagnosed with liver problems than children who did not receive a hepatitis B vaccine (95% CI of 1.07 to 8.05).[3] When considering only the children with vaccination records, the odds ratio in the vaccinated group increased to 13.08 times that of the group that had not been vaccinated

Hepatitis B Vaccine and Liver Problems in U.S. Children Less than 6 Years Old

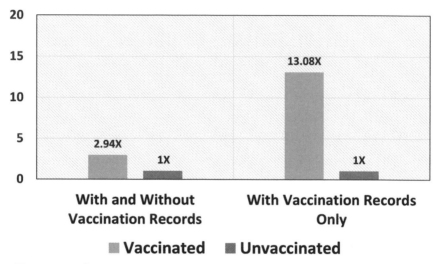

Figure 9.1—Odds ratios for liver problems in children ages 0 to 5 years who received at least one hepatitis B vaccine versus children unvaccinated against hepatitis B (Fisher et al. 1999).

against hepatitis B (95% CI of 2.66 to 64.39).[4] Both reported results were statistically significant.[5] The difference between the results "with and without vaccination records" and "with vaccination records only" may be due to individuals counted in the "unvaccinated" group who were vaccinated but didn't possess vaccination records.

Figure 9.2 shows results from the paper "Immunization with Hepatitis B Vaccine Accelerates SLE-Like Disease in a Murine Model," published in the *Journal of Autoimmunity* in 2014.[6] The lead author is Dr. Nancy Agmon-Levin, affiliated with the Zabludowicz Center for Autoimmune Diseases at the Sheba Medical Center in Tel-Hashomer, Israel. Dr. Yehuda Shoenfeld, the Incumbent of the

Immunization with Hepatitis B Vaccine Accelerates SLE-Like Disease in a Murine Model

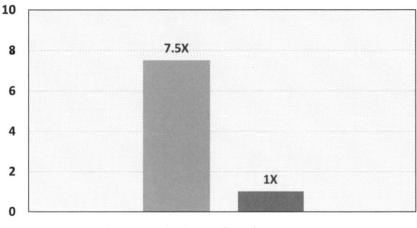

Protein in Urine of Mice Receiving the Hepatitis B Vaccine versus Saline Control

Figure 9.2—Protein in the urine (proteinuria) of female mice injected with hepatitis B vaccine compared to female mice injected with phosphate-buffered saline (Agmon-Levin et al. 2014).

Laura Schwarz-Kip, chair for autoimmunity at Tel Aviv University in Israel, is the corresponding author and is considered one of the world's foremost authorities on autoimmunity. In this study, researchers injected female mice with 0.4 milliliters of either Engerix® hepatitis B vaccine or phosphate-buffered saline at ages 8 and 12 weeks. Phosphate-buffered saline is inert and provides an appropriate placebo control. Investigators measured protein in the urine (proteinuria) as an indicator of kidney disease. Accordingly, protein levels in the urine of vaccinated female mice were 7.5 times higher than in the mice receiving phosphate-buffered saline (p-value < 0.004).[7] Also, mice injected with Engerix® showed severe and advanced nephropathology (kidney disease) compared to mice receiving either phosphate-buffered saline or aluminum adjuvant alone.[8]

Figure 9.3 shows results from the paper "The Timing of Pediatric Immunization and the Risk of Insulin-Dependent Diabetes Mellitus," published in *Infectious Diseases in Clinical Practice* in 1997.[9] The lead author is Dr. John B. Classen, affiliated with Classen Immunotherapies in Baltimore, Maryland. The incidence of type 1 diabetes in children living in Christchurch, New Zealand, rose from 11.2 per 100,000 (average between 1982 and 1987) to 18.1 per 100,000 (average between 1989 and 1991) after the introduction of the hepatitis B vaccination in 1988

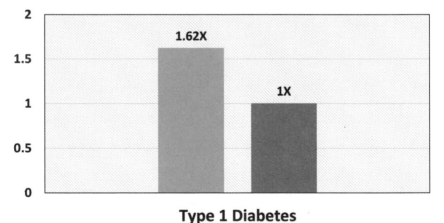

The Timing of Pediatric Immunization and the Risk of Insulin-Dependent Diabetes Mellitus

Incidence of Type 1 Diabetes in New Zealand Children Before and After the Introduction of the Hepatitis B Vaccine

Figure 9.3—Increase in type 1 diabetes incidence in New Zealand children with the introduction of the hepatitis B vaccine to the childhood vaccination schedule (Classen et al. 1997).

(p-value = 0.0008).[10] Over 70% of children under 16 were vaccinated within the first few years of the program.[11]

Figure 9.4 shows results from the paper "Recombinant Hepatitis B Vaccine and the Risk of Multiple Sclerosis: A Prospective Study," published in *Neurology* in 2004.[12] The lead author is Dr. Miguel A. Hernan, affiliated with the Department of Epidemiology at the Harvard School of Public Health in Boston, Massachusetts. Within the population of the United Kingdom's General Practice Research Database (GPRD), which included over 3 million patients, patients receiving a hepatitis B vaccine in the previous three years were 3.1 times more likely to receive a diagnosis of multiple sclerosis compared

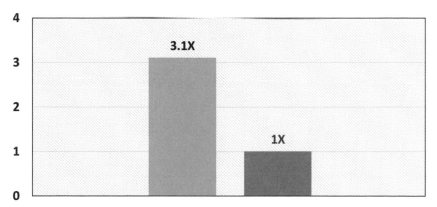

Figure 9.4—Multiple sclerosis incidence in patients receiving the hepatitis B vaccine versus those not receiving the hepatitis B vaccine (Hernan et al. 2004).

to patients who had not received a hepatitis B vaccine in the previous three years.[13] The difference in incidence was statistically significant, with a 95% confidence interval of 1.5 to 6.3.[14]

Figure 9.5 shows results from the paper "Hepatitis B Immunogenicity after a Primary Vaccination Course Associated with Asthma, Allergic Rhinitis, and Allergen Sensitization," published in the journal *Pediatric Allergy and Immunology* in 2018.[15] Dr. Dong Keon Yon, affiliated with the Department of Pediatrics at the CHA Bundang Medical Center in the CHA University School of Medicine in Seongnam, Korea, is

Hepatitis B Immunogenicity after a Primary Vaccination Course Associated with Asthma, Allergic Rhinitis, and Allergen Sensitization

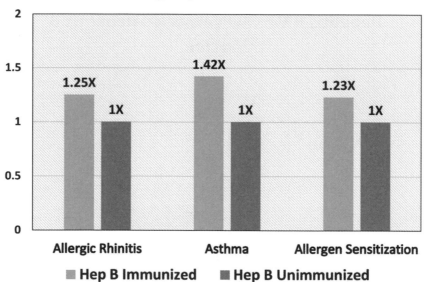

Figure 9.5—Odds ratios for allergic rhinitis, asthma, and allergen sensitization in all children receiving the hepatitis B vaccine. Those children who seroconverted to produce hepatitis B antibodies are compared to children who did not (Yon et al. 2018).

the study's lead author. Out of 3,176 Korean 12-year-old children who received the three-shot hepatitis B vaccine series as infants, 976 children were currently making antibodies to hepatitis B surface antigen, and 2,200 children were not.[16] Children who were hepatitis B surface antigen antibody positive showed a greater incidence of asthma (9.7% to 7.0%, p-value = 0.009), allergic rhinitis (33.3% vs. 28.8%, p-value = 0.013), and allergen sensitization (59.2% vs. 54.5%, p-value = 0.014) compared to vaccinated children who were antibody negative.[17] This study also demonstrates the waning immunity associated with hepatitis B vaccination in infancy as only 30.7% of those vaccinated were making hepatitis B-specific antibodies at age 12.[18]

Figure 9.6 shows our analysis of SIDS cases reported after the hepatitis B, *Haemophilus influenzae* B, diphtheria-tetanus-acellular

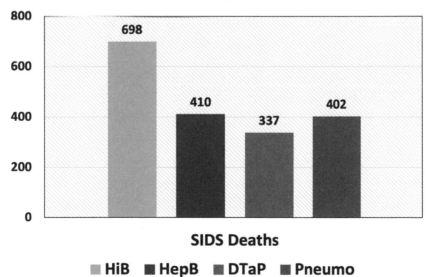

Figure 9.6—Number of SIDS deaths reported to the CDC Vaccine Adverse Events Reporting System (VAERS) database through June 16, 2023.

pertussis, and pneumonia vaccines.[19] VAERS has implicated hepatitis B vaccination in 410 SIDS deaths.[20] Many of these vaccines are given simultaneously, and some of the above reports were assigned to multiple vaccines.

Penina Haber (of the Immunization Safety Office of the Centers for Disease Control) and coworkers completed a brief survey of VAERS reports associated with the hepatitis B vaccine that included a significant treatment of infants receiving the vaccine alone or within a combined (multivalent) vaccine.[21] Overall, researchers reported 10,291 adverse event reports in children less than two years of age who received the hepatitis B vaccine over an 11-year period from Jan. 2005 to December 2015.[22] This included 197 reports of sudden infant death syndrome (SIDS).[23] From this, the study authors concluded, "Review [*sic*] the current US licensed HepB vaccines administered alone or in combination with other vaccines did not reveal new or unexpected safety concerns."[24] However, the study provided no basis for comparison regarding SIDS deaths. In a similar study involving vaccine adverse event surveillance in South Korea, the highest proportion of sudden death in infants following vaccination was related to the hepatitis B vaccine.[25]

Summary

Table 9.1 shows a summary of the five publications highlighted in Chapter 9 as well as an unpublished analysis of SIDS deaths associated with the hepatitis B vaccine from VAERS.[26, 27, 28, 29, 30, 31] There are other reports of adverse events associated with the hepatitis B vaccine (for example, Agmon-Levin et al. regarding chronic fatigue syndrome and fibromyalgia).[32] However, the publications highlighted in this chapter specifically compare vaccinated and unvaccinated populations. Additional hepatitis B vaccinated versus unvaccinated studies are also featured in Chapter 3 on thimerosal-containing vaccines.

	Fisher et al. 1999	Agmon-Levin et al. 2014	Classen et al. 1997	Hernan et al. 2004	Yon et al. 2013	VAERS analysis 2023
Liver Problems	✓					
Proteinuria in mice		✓				
Type 1 Diabetes			✓			
Multiple Sclerosis				✓		
Allergic Rhinitis					✓	
Asthma					✓	
Allergen Sensitization					✓	
SIDS						✓

Table 9.1—Summary of results comparing health outcomes of patients exposed to hepatitis B vaccines. Significantly higher odds ratios, relative risks or incidences are denoted by a ✓.

CHAPTER 10

COVID-19 Vaccines

The FDA authorized Pfizer's BNT162b2 COVID-19 vaccine under Emergency Use Authorization (EUA) for the US starting December 10, 2020. Other COVID-19 vaccines distributed in the US under EUA include the Moderna mRNA-1273 vaccine, the Johnson & Johnson Janssen vaccine, and the Novavax Nuvaxovid and Covovax vaccines. Full FDA approval was given to Pfizer (Comirnaty) and Moderna (Spikevax) vaccines. Pfizer and Moderna vaccines are based on mRNA technology, Novavax vaccines are based on recombinant protein technology, and the Johnson & Johnson vaccine is based on human adenovirus technology. As of May 7, 2023, the Johnson & Johnson Janssen vaccine is no longer available in the US. In Europe, the Oxford-AstraZeneca AZD1222 vaccine is based on the modified chimpanzee adenovirus ChAdOx1; and in China, the Sinovac CoronaVac vaccine is an inactivated virus vaccine. Many researchers have published studies investigating links between different types of COVID-19 vaccines and serious adverse events including myocarditis, pericarditis, blood-clotting disorders, shingles, hearing loss, hospitalizations, and death. This chapter presents studies where investigators directly compared vaccinated individuals to unvaccinated controls.

Bell's Palsy Adverse Events

Figure 10.1 shows results from the paper "Facial Nerve Palsy following the Administration of COVID-19 mRNA Vaccines: Analysis of a Self-Reporting Database," published in the *International Journal of Infectious Diseases* in 2021.[1] The lead author, Dr. Kenichiro Sato, is affiliated with the Department of Neurology in the Graduate School of Medicine at the University of Tokyo in Japan. Most often, practitioners reported the onset of facial nerve palsy three to four days following mRNA vaccination. Patients who received Pfizer's BNT162b2 vaccine showed the highest incidence of Bell's palsy compared to all other vaccines reported in the VAERS database (95% CI of 1.65 to 2.06).[2] Bell's palsy is a neurological disorder that causes paralysis or weakness on one side of the face.[3] Facial paralysis can vary from patient to patient and can be mild or severe. Patients typically recover some or

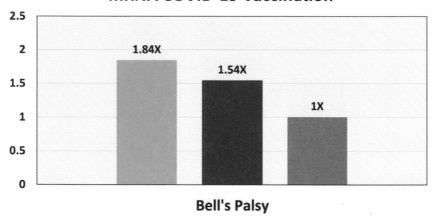

Figure 10.1—Odds ratios for Bell's palsy after the Pfizer BNT162b2 or Moderna mRNA-1273 vaccines compared to unvaccinated controls based on VAERS reports (Sato et al. 2021).

all their facial functions within a few weeks to six months. However, facial weakness and paralysis can be permanent.

Figure 10.2 shows results from the paper "Bell's Palsy following Vaccination with mRNA (BNT162b2) and Inactivated (CoronaVac) SARS-CoV-2 Vaccines: A Case Series and Nested Case-Control Study" in the journal *Lancet Infectious Diseases* in 2022.[4] The lead author, Dr. Eric Yuk Fai Wan, is affiliated with the Centre for Safe Medication Practice and Research, Department of Pharmacology and Pharmacy, Li Ka Shing Faculty of Medicine at the University of Hong Kong in China. This study uses patient data from the Hong Kong COVID-19 Vaccine Adverse Event Online Reporting system. Patients receiving Pfizer's BNT162b2 and Sinovac's CoronaVac vaccines had a higher risk for Bell's palsy than unvaccinated patients, with odds ratios of 1.75 and 2.38 and 95% confidence intervals of 0.886 to 3.477 and 1.415 to 4.002, respectively.[5]

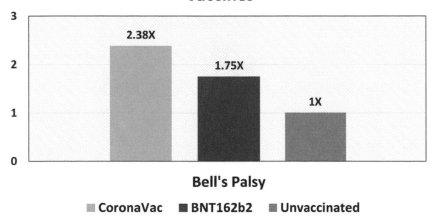

Figure 10.2—Odds ratios for Bell's palsy following the Pfizer BNT162b2 and Sinovac CoronaVac COVID-19 vaccines compared to unvaccinated individuals (Wan et al. 2022).

Figure 10.3 shows results from "Association between Vaccination with the BNT162b2 mRNA COVID-19 Vaccine and Bell's Palsy: A Population-Based Study," published in the *Lancet Regional Health–Europe* journal in 2021.[6] The lead author, Dr. Rana Shibli, MD, is affiliated with the Department of Community Medicine and Epidemiology, Lady Davis Carmel Medical Center, Haifa, Israel. This retrospective cohort study retrieved data on BNT162b2 mRNA (Pfizer) COVID-19 vaccination from December 2020 through April 2021 and the incidence of Bell's palsy from the database of the largest health-care provider in Israel, which included over 2.5 million vaccine recipients. The number of observed cases of Bell's palsy (designated by ICD [International Classification of Diseases] medical coding and filling a prescription of prednisone within two weeks after diagnosis) that occurred within 21 days after the first vaccine dose and within 30 days after the second vaccine dose were compared to expected cases, based

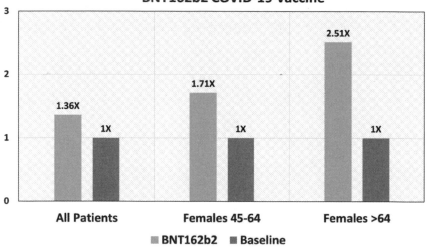

Figure 10.3 - Increased risk of Bell's palsy within 21 days of receiving the first dose of Pfizer BNT162b2 COVID-19 vaccine compared to the number of expected cases based on 2019 rates (Shibli et al. 2021).

on 2019 rates.[7] The first vaccine dose was associated with an increased risk of Bell's palsy with a standardized incidence ratio (SIR) of 1.36 (95% CI of 1.14 to 1.61).[8] Females aged 45–64 showed a higher SIR of 1.71 (95% CI of 1.10 to 2.54), and females aged ≥65 showed an SIR of 2.51 (95% CI of 1.65 to 3.68).[9] SIR is similar to the relative risk or risk ratio, where the incidence of the vaccinated group is compared to the incidence of the unvaccinated control group.

Figure 10.4 shows results from "Messenger RNA Coronavirus Disease 2019 (COVID-19) Vaccination with BNT162b2 Increased Risk of Bell's Palsy: A Nested Case-Control and Self-Controlled Case Series Study," published in the journal *Clinical Infectious Diseases* in 2023.[10] The lead author, Dr. Eric Yuk Fai Wan, is affiliated with the Centre for Safe Medication Practice and Research, Department of Pharmacology and Pharmacy, Li Ka Shing Faculty of Medicine at the University of

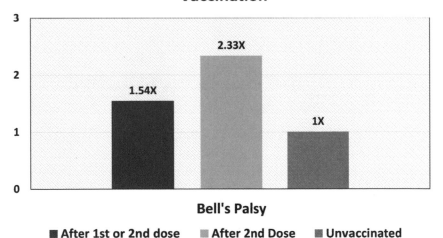

Figure 10.4 - Increased odds of hospitalization with Bell's palsy diagnosis in an inpatient setting within 28 days of vaccination with Pfizer BNT162b2 (Wan et al. 2023).

Hong Kong in China. This self-controlled case-control study used data from population-based electronic health records in individuals 16 years or older in Hong Kong to assess the diagnosis of Bell's palsy in an inpatient setting within 28 days of Pfizer BNT162b2 vaccination between March 2021 and July 2021. Vaccination with Pfizer BNT162b2 (first or second dose) yielded an increased odds of Bell's palsy diagnosis of 1.543 (95% CI of 1.123 to 2.121).[11] In addition, an increased odds of Bell's palsy of 2.325 was observed during the first 14 days after the second dose of BNT162b2 (95% CI of 1.414 to 3.821).[12]

Cardiac Adverse Events

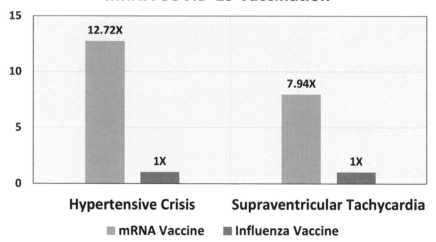

Figure 10.5—Risk of hypertensive crisis or supraventricular tachycardia among recipients of mRNA COVID-19 vaccines versus patients receiving the influenza vaccine (Kim et al. 2021).

Figure 10.5 shows results from the paper "Comparative Safety of mRNA COVID-19 Vaccines to Influenza Vaccines: A Pharmacovigilance Analysis Using WHO International Database," published in the *Journal of Medical Virology in 2021*.[13] The study's lead author, Dr. Min Seo Kim, is affiliated with the College of Medicine at Korea University in Seoul, South Korea. In this study, investigators compared cardiac adverse events from the mRNA COVID-19 vaccine to those from influenza vaccines using the WHO VigiBase for adverse events. Overall, individuals receiving COVID-19 mRNA vaccines showed a 12.72 times higher incidence of cardiac hypertensive crisis (95% CI of 2.47 to 65.54) and a 7.94 times higher incidence of supraventricular tachycardia (95% CI of 2.62 to 24.00) than those receiving the influenza vaccine.[14] Odds ratios were based on the relative incidence of each adverse event per number of each type of vaccine distributed.

Myocarditis and Pericarditis Adverse Events

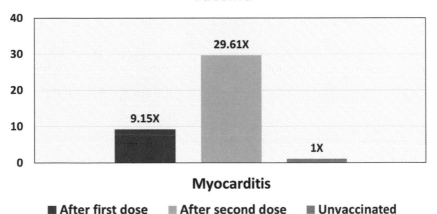

Figure 10.6—Increased risk of myocarditis in adolescents within 28 days of the first and second dose of the Pfizer BNT162b2 COVID-19 vaccine (Lai et al. 2022).

Figure 10.6 shows results from the paper "Adverse Events of Special Interest Following the Use of BNT162b2 in Adolescents: A Population-Based Retrospective Cohort Study," published in the journal *Emerging Microbes & Infections* in 2022.[15] The lead author,

Dr. Francisco Tsz Tsun Lai, is affiliated with the Centre for Safe Medication Practice and Research, Department of Pharmacology and Pharmacy, Li Ka Shing Faculty of Medicine, University of Hong Kong in China. In this study, adolescents ages 12 through 18 in Hong Kong who received the first dose of the Pfizer BNT162b2 vaccine had a 9.15 times greater risk of myocarditis compared to unvaccinated adolescents (95% CI of 1.14 to 73.16).[16] Those who received the second dose had a 29.61 times greater risk of myocarditis compared to unvaccinated adolescents (95% CI of 4.04 to 217.07).[17] Investigators assessed risks within 28 days of vaccination. In addition, after their second dose of the Pfizer BNT162b2 vaccine, vaccinated adolescents had a 2.06 times greater risk of sleep disturbances/disorders compared to unvaccinated adolescents (95% CI of 1.01 to 4.24).[18]

Myocarditis is a severe illness, indicating damage to the myocardium (heart muscle). Individuals at the highest risk include young adult males, although females may also contract myocarditis. Almost 20% of all sudden deaths in young people are due to myocarditis.[19] The survival rate for myocarditis is 80% after one year and 50% after five years.[20]

Figure 10.7 shows results from the paper "SARS-CoV-2 Vaccination and Myocarditis in a Nordic Cohort Study of 23 Million Residents," published in the journal *JAMA Cardiology* in 2022.[21] The lead author, Dr. Oystein Karlstad, is affiliated with the Department of Chronic Diseases at the Norwegian Institute of Public Health in Oslo, Norway. Participants in the study included 23,122,522 Nordic country residents ages 12 and older. Researchers observed the highest risk in males between 16 and 24 years of age after receiving the second Moderna mRNA-1273 (Incident Rate Ratio of 13.83 and a 95% CI of 8.08 to 23.68) or Pfizer BNT162b2 (Incident Rate Ratio of 5.31 and a 95% CI of 3.68 to 7.68) mRNA vaccine.[22]

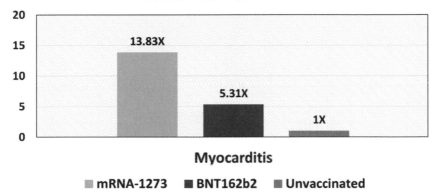

Figure 10.7—Increased risk for myocarditis in males aged 16 to 24 years following the second Pfizer BNT162b2 and Moderna mRNA-1273 COVID-19 vaccines (Karlstad et al. 2022).

Figure 10.8 shows results from the paper "Risk of Myocarditis after Sequential Doses of COVID-19 Vaccines and SARS-CoV-2 Infection by Age and Sex," published in the journal *Circulation* in 2022.[23] The lead author, Dr. Martina Patone, is affiliated with the Nuffield Department of Primary Health Care Services in Oxford, England. Researchers considered individuals in England aged 13 and older. This was a self-controlled study, meaning that researchers compared participants for disease incidence before and after COVID-19 vaccination. Men receiving the second dose of Moderna mRNA-1273 vaccine showed the highest levels of myocarditis with a relative risk of 14.98 and a 95% CI of 8.61 to 26.07.[24]

Figure 10.9 shows results from the paper "Acute Myocarditis following a Third Dose of COVID-19 mRNA Vaccination in Adults," published in the *International Journal of Cardiology* in 2022.[25] The lead author, Dr. Anthony Simone, is affiliated with the Department of Cardiology at the Kaiser Permanente Los Angeles Medical Center

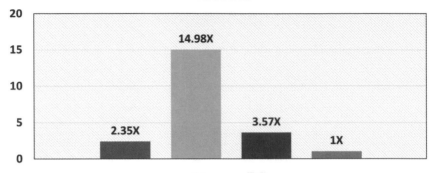

Figure 10.8—Risk of myocarditis in men who received the first, second, or third Moderna mRNA-1273 COVID-19 vaccine compared to unvaccinated men (Patone et al. 2022).

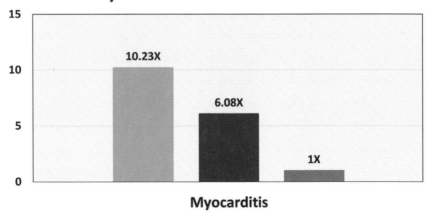

Figure 10.9—Increased risk of myocarditis within seven days of mRNA COVID-19 vaccination compared to a baseline period (Simone et al. 2022).

in California. This study included all Kaiser Permanente Southern California patients who received one to three doses of mRNA COVID-19 vaccines between December 14, 2020, and February 18, 2022. The risk of myocarditis within seven days of the second vaccine was 10.23 times higher than in the baseline period (p-value < 0.0001 and a 95% CI of 6.09 to 16.4).[26] The risk of myocarditis within seven days of the third vaccine (booster) was 6.08 times higher than in the baseline period (p-value < 0.0003 and a 95% CI 2.34 to 13.3).[27] The baseline period was 365 days, specified within two years prior to the vaccination date. No statistically significant risk was associated with the first dose of mRNA vaccine received in this study.

Figure 10.10 shows results from the paper "Carditis After COVID-19 Vaccination with a Messenger RNA Vaccine and an Inactivated Virus Vaccine: A Case-Control Study," published in *Annals of Internal Medicine* in 2022.[28] The colead authors, Dr. Francisco Tsz Tsun Lai,

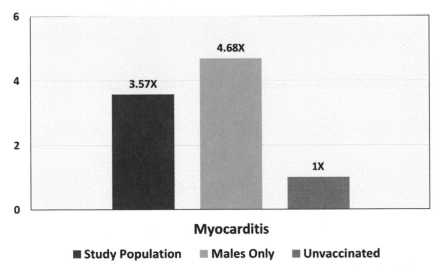

Figure 10.10 - Increased odds of carditis in hospital patients after the Pfizer BNT162b2 COVID-19 vaccine compared to unvaccinated patients (Lai et al. 2022).

PhD, and Dr. Xue Li, PhD, are affiliated with the Centre for Safe Medication Practice and Research, Department of Pharmacology and Pharmacy, Li Ka Shing Faculty of Medicine, The University of Hong Kong, and Laboratory of Data Discovery for Health (D24H), Hong Kong Science Park, Hong Kong Science and Technology Park, Hong Kong Special Administrative Region, China.

This case-control study of hospitalized patients ages 12 and older in Hong Kong from February to August 2021 assessed 160 patients with carditis and elevated troponin levels and 1,533 control patients. Multivariable analyses controlling for cardiovascular disease risk factors showed that recipients of the Pfizer BNT162b2 vaccine had 3.57 times greater odds of carditis than unvaccinated patients (95% CI of 1.93 to 6.60).[29] For male vaccine recipients, the odds were 4.68 times greater (95% CI of 2.25 to 9.71).[30] In addition, the risk was higher after the second dose of BNT162b2 than the first.[31]

Figure 10.11 shows results from the paper "Myocarditis after BNT162b2 mRNA Vaccine against Covid-19 in Israel," published in The *New England Journal of Medicine* in 2021.[32] The lead author, Dr. Dror Mevorach, MD, is affiliated with the Department of Internal Medicine B, Division of Immunology–Rheumatology, and Wohl Institute for Translational Medicine, Hadassah Medical Center in Israel. In this retrospective cohort study of Israeli Ministry of Health data, the incidence of myocarditis within 30 days after the second dose of the Pfizer BNT162b2 mRNA vaccine was 2.35 times higher than in unvaccinated people (95% CI of 1.10 to 5.02). The incidence ratio was highest in male recipients between the ages of 16 and 19 years at 8.96 cases per 10,857 (95% CI of 4.50 to 17.83) or roughly one in 1,000.[33] The researchers determined that the rate of myocarditis in the general unvaccinated population was one in 10,857.[34]

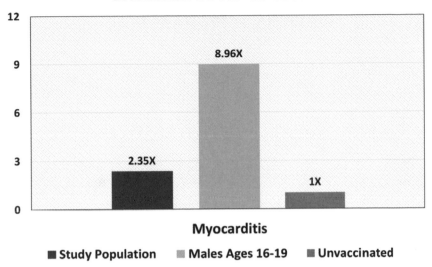

Increased Risk of Myocarditis After the Pfizer BNT162b2 COVID-19 Vaccine

Myocarditis

■ Study Population ■ Males Ages 16-19 ■ Unvaccinated

Figure 10.11—Increased risk of myocarditis within 30 days after the second dose of the Pfizer BNT162b2 mRNA vaccine for the entire study population and for male recipients between the ages of 16 and 19 years, compared to unvaccinated individuals (Mevorach et al. 2021).

Figure 10.12 shows results from the paper "Postmarketing Active Surveillance of Myocarditis and Pericarditis Following Vaccination with COVID-19 mRNA Vaccines in Persons Aged 12 to 39 years in Italy: A Multi-Database, Self-Controlled Case Series Study," published in the journal *PLOS Medicine* in 2022.[35] The lead author, Dr. Marco Massari, is affiliated with the National Centre for Drug Research and Evaluation at the Istituto Superiore di Sanità in Rome, Italy. In this study, males receiving the first or second dose of mRNA-1273 had approximately a twelve times greater risk of myocarditis (95% CI of 4.09 to 36.83) or pericarditis (95% CI of 3.88 to 36.53) within seven days of vaccination compared to the baseline period of December 27, 2020, through September 30, 2021, excluding the interval of 0 to 21 days after the first or second vaccine.[36]

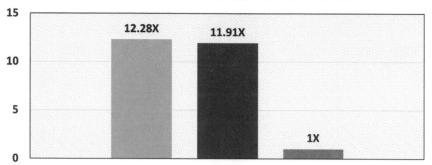

Figure 10.12—Increased risk of myocarditis or pericarditis in males aged 12 to 39 years within seven days of receiving the first or second dose of the Moderna mRNA-1273 COVID-19 vaccine (Massari et al. 2022).

Figure 10.13 shows results from the study "Risk of Myocarditis and Pericarditis Following BNT162b2 and mRNA-1273 COVID-19 Vaccination," published in the journal *Vaccine* in 2022.[37] The lead author, Dr. Kristin Goddard, is affiliated with the Kaiser Permanente Vaccine Study Center, Kaiser Permanente Northern California, in Oakland, California. Coauthors Dr. Eric Weintraub, Dr. Tom Shimabukuro, and Dr. Matthew Oster are affiliated with the Immunization Safety Office of the CDC in Atlanta, Georgia. Study participants from eight integrated healthcare delivery systems in the CDC's Vaccine Safety Datalink showed a significantly greater risk of myocarditis or pericarditis within seven days of receiving the first or second dose of the Pfizer (p-value = 0.044 and a 95% CI of 1.03 to 8.33 and p-value < 0.001 and a 95% CI of 6.45 to 34.85, respectively) or Moderna (p-value = 0.031 and a 95% CI of 1.12 to 11.07 and p-value < 0.001 and a 95% CI of 6.73 to 64.94, respectively) COVID-19 vaccines compared to participants within the baseline period of the study from December 14, 2020 to January 15, 2022, excluding the 0- to 7-day windows after vaccination.[38]

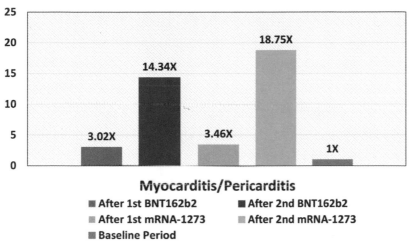

Figure 10.13—Increased risk of myocarditis or pericarditis in persons aged 18 to 39 years within 7 days of receiving the first or the second dose of the Pfizer BNT162b2 or Moderna mRNA-1273 COVID-19 vaccine (Goddard et al. 2022).

Thrombocytopenia and Thrombosis

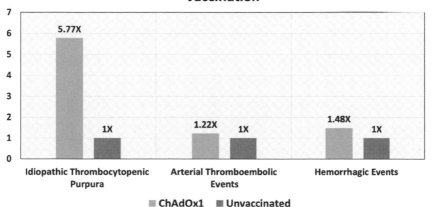

Figure 10.14—Increased Risk of thrombocytopenic, thromboembolic, and hemorrhagic events following Oxford-AstraZeneca (ChAdOx1) COVID-19 vaccination, compared to unvaccinated people (Simpson et al. 2021).

Figure 10.14 shows results from "First Dose ChAdOx1 and BNT162b2 COVID-19 Vaccines and Thrombocytopenic, Thromboembolic and Hemorrhagic Events in Scotland," published in the journal *Nature Medicine* in 2021.[39] The lead author, Dr. Colin R. Simpson, PhD, is affiliated with the School of Health, Wellington Faculty of Health, Victoria University of Wellington in New Zealand, and the Usher Institute, University of Edinburgh, in the UK. This national prospective cohort study from Scotland included over 2.5 million people over the age of 18 vaccinated between December 2020 and April 2021. Oxford-AstraZeneca (ChAdOx1) vaccination was associated with a 5.77 times increased risk of idiopathic thrombocytopenic purpura (autoimmune clotting disorder) 0 to 27 days after vaccination (95% CI of 2.41 to 13.83).[40] Oxford-AstraZeneca vaccination was also associated with an increased risk of arterial thromboembolic events (arterial blood clots) with an adjusted relative risk of 1.22, 0 to 27 days after vaccination (95% CI of 1.12 to 1.34), and hemorrhagic events (excessive bleeding) with an adjusted relative risk of 1.48, 0 to 27 days after vaccination (95% CI of 1.12 to 1.96).[41]

Figure 10.15 shows results from the paper "Analysis of Thromboembolic and Thrombocytopenic Events after the AZD1222, BNT162b2, and mRNA-1273 in Three Nordic Countries," published in the journal *JAMA Network Open* in 2022.[42] The study's lead author, Dr. Jacob Dag Berild, is affiliated with the Department of Infection Control and Vaccines at the Norwegian Institute of Public Health in Oslo, Norway. In this study, investigators used hospital registries from Norway, Finland, and Denmark to measure the incidence of thrombocytopenia and cerebral venous thrombosis within 28 days following several available COVID-19 vaccines. Thrombocytopenia is a deficiency of platelets in circulating blood and can lead to spontaneous bleeding. Cerebral venous thrombosis occurs when a blood clot blocks blood flow away from the brain and can be a cause of

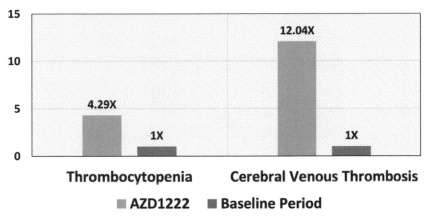

Figure 10.15—Increased risk of thrombocytopenia and cerebral venous thrombosis within 28 days following the AstraZeneca AZD1222 (ChAdOx1) vaccine compared to the baseline period (Berild et al. 2022).

stroke. Researchers observed the highest risks for patients receiving the AstraZeneca COVID-19 vaccine, with a 4.29 times higher risk of thrombocytopenia (95% CI of 2.96 to 6.20) and a 12.04 times higher risk of cerebral venous thrombosis (95% CI of 5.37 to 26.99).[43] The baseline period for comparison was between January 1, 2020, and May 16, 2021, excluding the 28-day window after vaccination for each patient considered.

Shingles

Figure 10.16 shows results from the paper "Herpes Zoster Related Hospitalization after Inactivated (CoronaVac) and mRNA (BNT162b2) SARS-CoV-2 Vaccination: A Self-Controlled Case Series and Nested Case-Control Study," published in the journal *Lancet Regional Health— Western Pacific* in 2022.[44] The lead author is Dr. Eric Yuk Fai Wan. Patients in this study who received the Pfizer BNT162b2 vaccine were more than five times more likely to contract shingles in the intervals

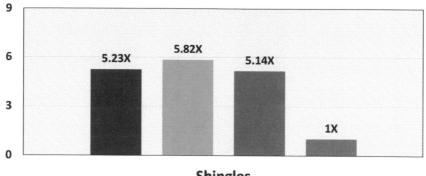

Figure 10.16—Increased risk of shingles (herpes zoster) in two-week intervals after the Pfizer BNT162b2 vaccine compared to a baseline period (Wan et al. 2022).

from 0 to 13 (95% CI of 1.61 to 17.03) and 14 to 27 days from the first vaccine received (95% CI of 1.62 to 20.91) and 0 to 13 days after the second vaccine was received (95% CI of 1.29 to 20.47) than during the baseline period, which was at any time between February 23 and July 31, 2021, outside of the specified time frames from vaccination.[45] In addition, patients receiving the CoronaVac vaccine were 2.67 times more likely to contract shingles within 13 days of vaccination (95% CI of 1.08 to 6.59).[46] Shingles is a painful, sometimes serious condition resulting from the reactivation of the herpes zoster virus that causes chicken pox. Anyone who has had chicken pox, or the varicella vaccine, may be at risk of this reactivation when their immune system is compromised or suppressed.

Hearing Loss

Figure 10.17 shows results from the paper "Association between BNT162b2 Messenger RNA COVID-19 Vaccine and Risk of Sudden Sensorineural Hearing Loss," published in the journal *JAMA*

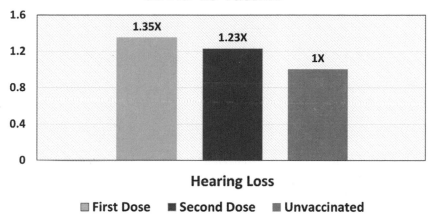

Figure 10.17—Increased risk of sudden sensorineural hearing loss following the Pfizer BNT162b2 COVID-19 vaccine (Yanir et al. 2022).

Otolaryngology–Head & Neck Surgery in 2022.[47] The lead author is Dr. Yoav Yanir from the Department of Otolaryngology-Head and Neck Surgery, Lady David Carmel Medical Center, Haifa, Israel. This is a population-based study of a large healthcare organization in Israel. Standardized incidence ratios (SIR) for sudden sensorineural hearing loss associated with the first and second doses of the Pfizer BNT162b2 were 1.35 (95% CI of 1.09 to 1.65) and 1.23 (95% CI of 0.98 to 1.53), respectively. Risks were greatest after the first dose in females aged 16 to 44 years (SIR 1.92, 95% CI 0.98 to 3.43) and in females older than 65 years (SIR 1.68, 95% CI 1.15 to 2.37) and after the second dose in males aged 16 to 44 years (SIR 2.45, 95% CI 1.36 to 4.07).[48] Patients with sudden sensorineural hearing loss can experience tinnitus. It can also lead to permanent hearing loss.

Adverse Events of COVID-19 Vaccines versus Influenza Vaccines

Figure 10.18 shows results from the paper "Frequency and Associations of Adverse Reactions to COVID-19 Vaccines Reported

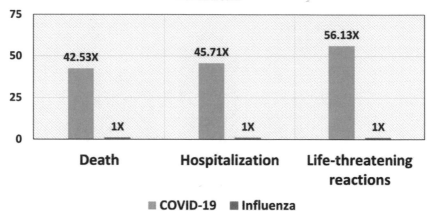

Figure 10.18—Increased risk of adverse events from COVID-19 vaccines versus influenza vaccines based on adverse event reports to the European Database of Suspected Adverse Drug Reactions (Montano 2022).

to Pharmacovigilance Systems in the European Union and the United States," published in the journal *Frontiers in Public Health* in 2022.[49] The paper's author, Dr. Diego Montano, is affiliated with the Department of Population-Based Medicine at the Institute of Health Sciences at the University of Tubingen in Germany. Dr. Montano compared adverse events reports from EudraVigilance (The European Database of Suspected Adverse Drug Reactions) for COVID-19 and influenza vaccines. These were normalized to the European Centers for Disease and Prevention (ECDC) estimates of the total number of each type of vaccine administered. Death (99% CI of 33.49 to 54.01), hospitalization (99% CI 41.26 to 50.65), and life-threatening reaction (99% CI of 44.51 to 70.78) reports per unit of COVID-19 vaccine given far eclipsed those for the influenza vaccine.[50] The author also reported significant relative risks of thrombosis, coagulation, and sexual organ reactions associated with COVID-19 vaccines.[51]

Various Adverse Events

Figure 10.19 shows results from the paper "Surveillance of COVID-19 Vaccine Safety among Elderly Persons Aged 65 Years and Older," published in *Vaccine* in 2023.[52] The lead author, Dr. Hui-Lee Wong, PhD, is affiliated with the Food and Drug Administration, Silver Spring, Maryland. This FDA-sponsored prospective study tracked US Medicaid claims data for over 30 million patients aged 65 and older from December 2020 through January 2022. Weekly sequential testing revealed four outcomes that met the threshold for a statistical signal following Pfizer BNT162b2 vaccination compared to pre-COVID-19 vaccine historic levels: pulmonary embolism (blood clot in the lungs), with a relative risk of 1.54 between 1 and 28 days post-vaccination; acute myocardial infarction (heart attack), with a relative risk of 1.42 between 1 and 28 days post-vaccination; disseminated intravascular coagulation (abnormal blood clotting throughout the body), with a relative risk of

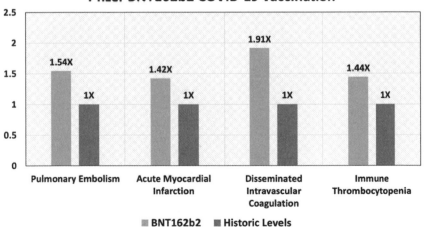

Figure 10.19—Increased risk of various adverse events among elderly persons aged 65 years and older following Pfizer BNT162b2 vaccination compared to an unvaccinated population prior to the pandemic (Wong et al., 2023).

1.91 between 1 and 28 days post-vaccination; and immune thrombocy-topenia (low blood platelets due to autoimmune attack), with a relative risk of 1.44 between 1 and 42 days post-vaccination.[53]

Serious Adverse Events

Figure 10.20 shows results from the paper "Serious Adverse Events of Special Interest following mRNA COVID-19 Vaccination in Randomized Trials in Adults," published in *Vaccine* in 2022.[54] The lead author, Dr. Joseph Fraiman, is affiliated with Thibodaux Regional Health System in Thibodaux, Louisiana. The corresponding author, Dr. Peter Doshi, is affiliated with the School of Pharmacy at the University of Maryland in Baltimore and is a senior editor of *BMJ*.

Using data from the phase III clinical trials for the Pfizer BNT162b2 and Moderna mRNA-1273 vaccines, investigators directly compared vaccine recipients to placebo control recipients. Overall, recipients of

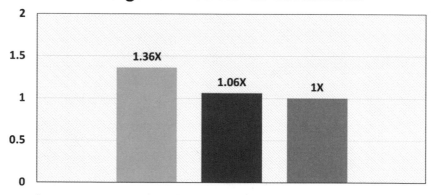

Figure 10.20—Risk ratios for serious adverse events following the Pfizer BNT162b2 or Moderna mRNA-1273 vaccines compared to unvaccinated controls (Fraiman et al. 2022).

either of the mRNA vaccines had a 1.16 times greater risk of serious adverse events compared to unvaccinated controls (95% CI of 0.97 to 1.39).[55] The result was marginally statistically significant.

The study authors also completed a risk-benefit assessment of both vaccines. They found that the Pfizer BNT162b2 vaccine showed an excess risk of serious adverse events of special interest of 10.1 per 10,000 vaccinees (95% CI of -0.4 to 20.6) while preventing COVID-19 hospitalizations in 2.3 per 10,000 vaccinees compared to the placebo group.[56] Likewise, the Moderna mRNA-1273 vaccine showed an excess risk of serious adverse events of special interest of 15.1 per 10,000 vaccinees (95% CI of -3.6 to 33.8) while preventing COVID-19 hospitalizations in 6.4 per 10,000 vaccinees compared to the placebo group.[57]

In this paper, a serious adverse event of special interest was defined as death, life-threatening at the time of the event; inpatient hospitalization, or prolongation of existing hospitalization; persistent or significant disability/incapacity; a congenital anomaly/birth defect; or a medically important event based on medical judgment.

Summary

Tables 10.1 and 10.2 show a summary of the results from the scientific literature where researchers compared COVID-19 vaccine recipients to unvaccinated controls. Investigators showed that myocarditis or pericarditis was significantly associated with COVID-19 vaccination in eight different studies identified.[58, 59, 60, 61, 62, 63, 64, 65] Researchers also demonstrated a significant association between Bell's palsy and COVID-19 vaccination in four studies as compared to those unvaccinated for COVID-19.[66, 67, 68, 69]

	Lai et al. 2022a	Lai et al. 2022b	Kim et al. 2021	Massari et al. 2022	Goddard et al. 2022	Wong et al. 2023	Mevorach et al. 2021	Patone et al. 2022	Simpson et al. 2021	Simone et al. 2022	Karlstad et al. 2022
Pulmonary Embolism						✓					
Myocarditis or Pericarditis	✓	✓		✓	✓		✓	✓		✓	✓
Hypertensive Crisis			✓								
Supraventricular Tachycardia				✓							
Myocardial Infarction	.					✓					
Disseminated Intravascular Coagulation						✓					
Immune Thrombocytopenia						✓					
Thrombocytopenic Purpura									✓		
Arterial Thrombolytic Events									✓		
Hemorrhagic Events									✓		

Table 10.1—Summary of results in comparing health outcomes of COVID-19 vaccinated versus unvaccinated individuals. Significantly higher odds ratios, relative risks, or incidences are denoted by a ✓.

	Wan et al. 2022	Cheng et al. 2021	Sato et al. 2021	Fraiman et al. 2022	Shibli et al. 2021	Montano 2022	Berild et al. 2022	Yanir et al. 2022	Wan et al. 2022	Wan et al. 2023	Khoupayeh et al. 2022
Thrombocytopenia							✓				
Cerebral Venous Thrombosis							✓				
Shingles	✓										
Hearing Loss								✓			
Bell's Palsy			✓		✓				✓	✓	
Adverse Events in General		✓		✓							✓
Life-Threatening Reactions						✓					
Hospitalization						✓					
Death						✓					

Table 10.2 - Summary of results in comparing health outcomes of COVID-19 vaccinated versus unvaccinated individuals. Significantly higher odds ratios, relative risks or incidences are denoted by a ✓.

CHAPTER 11

Vaccines in Pregnancy

Medical practitioners routinely give pregnant women flu, Tdap, and COVID-19 vaccines during any trimester of pregnancy. While occasionally a clinical trial participant may become pregnant during a study, the FDA has never deliberately tested these products on a single pregnant woman as a part of the approval process. In fact, up until 2020, the package insert for the Boostrix® Tdap vaccine stated, "There are no adequate and well-controlled studies of BOOSTRIX in pregnant women in the US."[1] Packet inserts for trivalent inactivated influenza (TIV) vaccines, Fluvirin® and Flublok®, which are recommended for pregnant women by the CDC, had similar wording. The Fluvirin® packet insert still warns, "Safety and Effectiveness of FLUVIRIN® have not been established in pregnant women . . ."[2] Similarly, for the FDA-approved version of the COVID-19 vaccine, Comirnaty®, which Pfizer manufactures, the package insert states, "[a]vailable data on COMIRNATY administered to pregnant women are insufficient to inform vaccine-associated risks in pregnancy."[3] The package insert for the FDA-approved Moderna COVID-19 vaccine, Spikevax, provides an identical disclosure.[4] Both inserts reference one small-scale animal reproductive toxicology study that showed no harm. However, Pfizer and Moderna did not conduct any clinical trials for pregnant women.

Yet the CDC recommended these COVID-19 vaccines to pregnant women without any safety testing or precautions.[5,6] Only after the recommendations were made and many millions of women received the vaccines did the manufacturers, CDC, or FDA even attempt to investigate safety in pregnant women by setting up registries to "monitor pregnancy outcomes" in women who received a COVID-19 vaccination.[7] This means every pregnant woman injected is an unwitting subject in a poorly administered experiment.

In this chapter, we investigate what the literature says about pregnancy outcomes in women who received influenza, Tdap, and/or COVID-19 vaccines during pregnancy versus those who did not. In addition, we examine studies that look at fertility outcomes associated with vaccination prior to conception.

Association between Influenza Infection and Vaccination during Pregnancy and Risk of Autism Spectrum Disorder

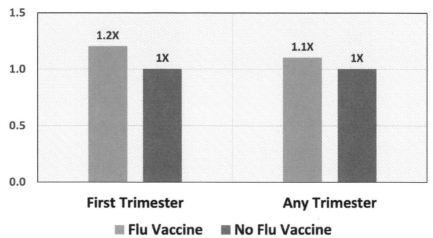

Figure 11.1 Hazard ratios for ASD incidence in the offspring of women receiving the flu vaccine in the first trimester and any trimester of pregnancy compared to the offspring of unvaccinated pregnant women (Zerbo et al. 2017).

Figure 11.1 shows results from the paper "Association between Influenza Infection and Vaccination during Pregnancy and Risk of Autism Spectrum Disorder," published in the journal *JAMA Pediatrics* in 2017.[8] The lead investigator, Dr. Ousseny Zerbo, is from the Division of Research at Kaiser Permanente in Oakland, California. Dr. Zerbo and his coauthors investigated the relationship between influenza infection and vaccination during pregnancy and autism spectrum disorder. They found that first-trimester influenza vaccination is associated with an increased risk of ASD, with a hazard ratio of 1.20 and a 95% confidence interval between 1.04 and 1.39.[9] Investigators also observed that flu vaccination in any trimester was associated with autism spectrum disorder with a hazard ratio of 1.10 and a 95% confidence interval between 1.00 and 1.21.[10]

After obtaining these statistically significant results, the investigators applied the Bonferroni correction, which is sometimes used when multiple statistical tests are completed on the same data sample. When statisticians make many comparisons within a single study, the "false positive" rate, or the probability of finding associations that are not valid, can increase when the comparisons made are independent or unrelated.[11] The Bonferroni correction adjusts for this. Erroneously using the correction, the authors raised the p-value from 0.01 to 0.1, which is above the threshold for the statistical significance (p-value less than 0.05). The researchers then claimed insignificant statistical relationships. This was countered by Dr. Alberto Donzelli[12] and Dr. Brian Hooker,[13] who wrote two separate letters to the editor of *JAMA Pediatrics* in 2017. They showed it was inappropriate for Dr. Zerbo[14] to apply any correction for multiple testing because all associations made in the study were highly interdependent and not independent, which is required for correction. For example, Dr. Zerbo's results for each trimester roll up into a total result, showing interdependency instead of independence. Therefore, any correction for a "false positive rate" would not apply.[15] Thus, the actual p-value for the analysis was 0.01, which is statistically significant.[16]

Trivalent Inactivated Influenza Vaccine and Spontaneous Abortion

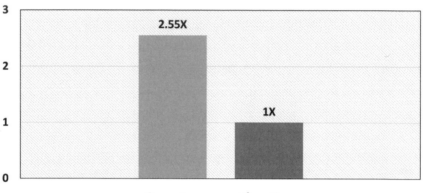

Figure 11.2—Odds ratio for spontaneous abortion when receiving the flu shot prior to conception versus unvaccinated women (Irving et al. 2013).

Figure 11.2 shows results from the study "Trivalent Inactivated Influenza Vaccine and Spontaneous Abortion," published in *Obstetrics & Gynecology* in 2013.[17] The study's lead author, Stephanie Irving, is from the Epidemiology Research Center at Marshfield Clinic in Marshfield, Wisconsin. A contributing author to the study is Dr. Frank DeStefano from the Immunization Safety Office of the CDC. Irving focused on the prenatal flu shot and the incidence of spontaneous abortion (SAB), the medical term for miscarriage, between five and sixteen weeks of gestation, when the vaccine was given during the first trimester of pregnancy. Investigators did not observe an increase in spontaneous abortions in women who were vaccinated prenatally compared to unvaccinated women. However, they found that women who received a flu shot before conception experienced a statistically significant increase. In this case, vaccinated women had 2.55 times greater odds of spontaneous abortion than unvaccinated women (p-value < 0.10 and a 95% CI of 0.86 to 6.33).[18] This result is marginally significant and deserves further analysis. In the study, 22 cases of spontaneous abortion and 11 controls received the influenza vaccine prior to conception, and a statistically significant proportion of the cases received the vaccine within seven days of conception.[19] Unlike in other flu shot studies[20, 21] Irving did not consider the effect of influenza vaccination in prior flu seasons.[22]

Comparison of VAERS Fetal-Loss Reports during Three Consecutive Influenza Seasons: Was There a Synergistic Fetal Toxicity Associated with the Two-Vaccine 2009/2010 Season?

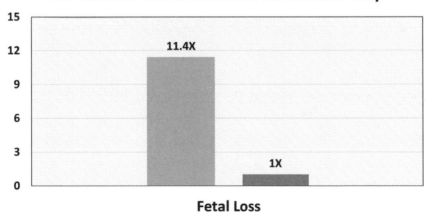

Figure 11.3—Relative risk for fetal loss in pregnant women receiving both the H1N1 and seasonal flu shots versus women receiving the seasonal flu shot only (Goldman 2013).

Figure 11.3 shows results from the paper "Comparison of VAERS Fetal-Loss Reports during Three Consecutive Influenza Seasons: Was There a Synergistic Fetal Toxicity Associated with the Two-Vaccine 2009/2010 Season?" published in the journal *Human and Experimental Toxicology* in 2013.[23] The study's author, Dr. Gary Goldman, is an independent computer scientist residing in Pearblossom, California. Goldman examined fetal-loss rates during three consecutive flu seasons. According to VAERS, the unadjusted fetal-loss rates from 2008 to 2009 were 6.8 per one million pregnancies, with a 95% confidence interval of 0.1 to 13.1.[24] From 2009 to 2010, rates rose to 77.8 per one million pregnancies with a 95% confidence interval of 66.3 to 89.4, and from 2010 to 2011, the rate was 12.6 with a 95% confidence interval of 7.2 to 18.0.[25] During the 2009 to 2010 season, where investigators saw a dramatic elevenfold increase in fetal loss, many pregnant women received two flu shots, the seasonal influenza vaccine, many of which contained 25 micrograms of mercury from thimerosal, and the H1N1 pandemic influenza vaccine, which also included 25 micrograms of mercury from thimerosal.[26]

In contrast, pregnant women received only the seasonal flu shot from 2008 to 2009, and in 2010 to 2011, they received a single "combination" vaccine. Also, from 2008 to 2009, only 11.3% of pregnant women received the seasonal flu shot, whereas from 2009 to 2010, a reported 43% received the H1N1 vaccine. From 2010 to 2011, 32% received the combination vaccine.[27] Goldman suggested that this elevenfold increase in fetal loss may be due to receiving an additional dose of thimerosal in the H1N1 shot.[28]

Figure 11.4 shows results from the paper "Association of Spontaneous Abortion with Receipt of Inactivated Influenza Vaccine Containing H1N1pdm09 in 2010–11 and 2011–12," published in the journal *Vaccine* in 2017.[29] The lead author, Dr. James Donahue, is senior epidemiologist at the Marshfield Clinic Research Institute

Association of Spontaneous Abortion with Receipt of Inactivated Influenza Vaccine Containing H1N1pdm09 in 2010–11 and 2011–12

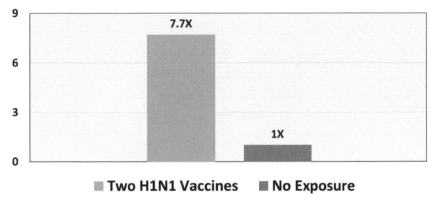

Odds Ratio for Miscarriage Within 28 Days of H1N1-Containing Influenza Vaccine in Women Receiving the Same Vaccine in the Previous Year

Figure 11.4—Odds ratio for miscarriage in women who received two H1N1 vaccines in consecutive years versus unvaccinated women (Donahue et al. 2017).

in Marshfield, Wisconsin. The Immunization Safety Office at the CDC collaborated with Marshfield Clinic researchers on this effort. In this study, women who received an H1N1 vaccine had an odds ratio of 2.0 for fetal loss within 28 days of vaccination in either of the two "flu seasons" 2010–2011 and 2011–2012, with a 95% confidence interval of 1.1 to 3.6, compared to pregnant women who did not receive the vaccine within the 28-day exposure window.[30] The odds ratio for women who received the pH1N1 vaccine during the previous season increased to 7.7 with a 95% confidence interval of 2.2 to 27.3.[31]

In a follow-up study, Donahue investigated the risk of spontaneous abortion from the seasonal flu shot.[32] The study authors saw no statistically significant effect. However, the study was underpowered due to the small cohort of women observed. Consistently, investigators used fewer than 100 case and control pairs for the combined analysis of three influenza seasons and derived some results for individual seasons from as few as 11 cases and control pairs. In addition, researchers completed a power analysis as a part of this study, which is an analysis of how much statistical power the study possesses to find an association. In this study, the *minimum* odds ratio that could be detected with statistical certainty was a 3.5.[33] The odds ratios reported in the results of this study were all below 2.0.[34] Thus, if this represented an actual increase in the miscarriage rate due to influenza vaccination, this study could not capture it, rendering the entire analysis meaningless.

Figure 11.5 shows results from the paper "Influenza Vaccination of Pregnant Women and Serious Adverse Events in the Offspring,"

Influenza Vaccination of Pregnant Women and Serious Adverse Events in the Offspring

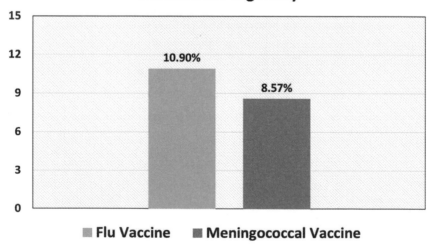

Figure 11.5—Incidence of severe adverse events after receiving the flu vaccine in pregnancy compared to the meningococcal vaccine (Donzelli et al. 2019a).

published in the *International Journal of Environmental Research and Public Health* in 2019.[35] Dr. Alberto Donzelli, the paper's author, is from the Scientific Committee of the Fondazione Allineare Sanità e Salute in Milan, Italy. In this paper, he reanalyzed data from four randomized controlled trials (RCTs) studying the maternal influenza vaccine. Donzelli showed that pregnant women receiving the flu shot had a higher incidence of serious adverse events (SAEs) than women who received a meningococcal vaccine. Within a RCT completed by Tapia et al. comparing a trivalent inactivated-virus influenza vaccine (test) versus a meningococcal vaccine (control),[36] Donzelli calculated total SAEs at 225 or 10.90% and 175 or 8.57% for the vaccinated and the "control" group, respectively.[37] From this, Dr. Donzelli obtained a relative risk of 1.27 in the flu shot group with a 95% CI of 1.05 to 1.53, which was statistically significant.[38] The number of vaccines given "needed to harm" a single individual was 42.98.[39] Recall that RCT is the "gold standard" of clinical studies, as was stated in Chapter 1.

Original RCT investigators buried these data in a supplement to their scientific paper. The paper's abstract reported a statistically significant increase in neonatal infection in the influenza-vaccinated group, with a p-value equaling 0.02.[40] However, as observed by Donzelli, the RCT study authors neglected to mention that they also found serious adverse events in addition to neonatal infections. These researchers also failed to point out that total spontaneous abortions were significantly higher in the influenza-vaccinated group.[41]

It is not evident why the original RCT investigators gave the control group the meningococcal vaccine that is not recommended for pregnant women rather than an inert saline placebo. Unfortunately, this choice guaranteed that the trial would not yield an accurate safety profile for trivalent inactivated-virus influenza vaccine in pregnant women. The fact that the trial's designers made that choice may indicate that they believed the number of serious adverse events would be unacceptably high compared to a true placebo control.

Figure 11.6 shows the results of the paper "Influenza Vaccination for All Pregnant Women? So Far the Less Biased Evidence Does Not Favor It," published in the journal *Human Vaccines & Immunotherapeutics* in 2019.[42] The author of this paper, Dr. Alberto Donzelli, continued to reanalyze the randomized controlled trial (RCT) completed by Tapia et al. Within this RCT, women in the flu vaccine group experienced 52 miscarriages versus women in the control group (the meningococcal vaccine), who experienced 37 miscarriages.[43] This gave a marginally significant relative risk of 1.39 for miscarriage in the influenza-vaccinated group (p-value = 0.122 and a 95% CI of 0.92 to 2.11).[44]

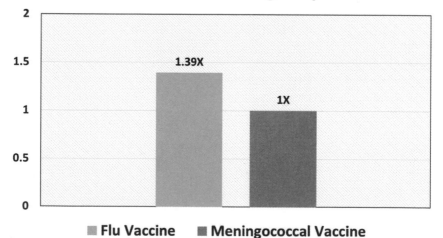

Figure 11.6—Incidence of spontaneous abortion in pregnant women receiving the flu vaccine versus the meningococcal vaccine (Donzelli et al. 2019b).

Figure 11.7 shows the results from the paper "Inflammatory Response to Trivalent Influenza Virus Vaccine among Pregnant Women," published in *Vaccine* in 2011.[45] The lead author is Dr. Lisa Christian, affiliated with the Department of Psychiatry, the Institute of Behavioral Medical Research, the Department of Psychology, and the Department of Obstetrics and Gynecology at The Ohio State University Medical Centers in Columbus. Adverse pregnancy outcomes, such as pre-eclampsia and preterm birth, are associated with higher levels of inflammation.[46] Accordingly, pregnant women receiving the trivalent inactivated-virus influenza vaccine showed increases in C-reactive protein with a p-value less than 0.05 and tumor necrosis factor α (TNF-α) with a p-value equaling 0.06 two days after vaccination.[47]

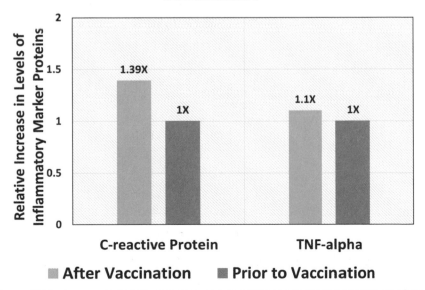

Inflammatory Response to Trivalent Influenza Virus Vaccine among Pregnant Women

Figure 11.7—Increases in inflammatory marker proteins in pregnant women after influenza vaccination compared to before vaccination (Christian et al. 2011).

C-reactive protein and TNF-α are inflammatory markers; high values indicate high levels of inflammation in the body. Elevated inflammatory markers are considered normal when the body fights an acute infection. Still, they may indicate a chronic condition, such as an autoimmune disease, if high levels of inflammation persist. The study authors noted that the observed increases were likely smaller than those associated with influenza infection.[48] However, the inflammatory parameters investigated in this study varied substantially from person to person.[49] This brings the risk-to-benefit ratio of influenza vaccination during pregnancy into question, especially on the individual level, given the potential severity of adverse outcomes associated with inflammation.

Figure 11.8 shows results from the paper "Prenatal and Infant Exposure to Thimerosal from Vaccines and Immunoglobulins and

Prenatal and Infant Exposure to Thimerosal from Vaccines and Immunoglobulins and Risk of Autism

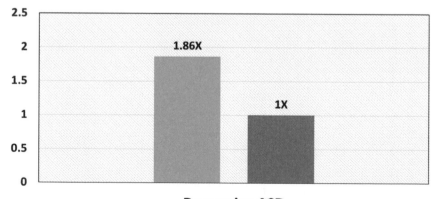

Figure 11.8—Odds ratio for regressive ASD from prenatal thimerosal exposure comparing pregnant women exposed to an average of 16 micrograms of thimerosal to pregnant women who were not exposed to thimerosal (Price et al. 2010).

Risk of Autism," published in *Pediatrics* in 2010.[50] Cristopher Price, an epidemiologist with Abt Associates in Cambridge, Massachusetts, is the lead author. Dr. Frank DeStefano, formerly the director of the Immunization Safety Office of the CDC in Atlanta, Georgia, is the corresponding author. The study authors investigated the effects of exposure in pregnancy to thimerosal from either influenza vaccines or immunoglobulin given to Rh-negative mothers (e.g., RhoGAM). They calculated a mean prenatal exposure in the study cohort of approximately two to three micrograms, as most women received neither the flu shot nor anti-rhoD immunoglobulin.[51]

The investigators used a difference of two standard deviations of exposure—or approximately 16.34 micrograms of mercury from thimerosal—as the threshold for the analysis.[52] Unfortunately, this is an artificial metric, given that the standard dose of mercury from thimerosal in a single flu shot is 25 micrograms.[53] Nonetheless, the study authors reported a marginally significant relationship between prenatal thimerosal exposure and regressive autism spectrum disorder with an odds ratio of 1.86 and a 95% confidence interval between 0.945 and 3.660.[54]

Notably, in the background studies leading to this publication, investigators ran six different variations of the model for prenatal thimerosal exposure and regressive autism spectrum disorder.[55] In two of the analyses, investigators found highly statistically significant results. In the remaining four analyses, researchers saw marginally statistically significant results. Unfortunately, the CDC only highlighted a marginally significant result and buried a highly significant result in the original study report, "Thimerosal and Autism" (by Abt Associates), on pages 194 and 197.[56] The CDC has not completed a follow-up study on this result and continues to deny a role between thimerosal exposure and autism.

Evaluation of the Association of Maternal Pertussis Vaccination with Obstetric Events and Birth Outcomes

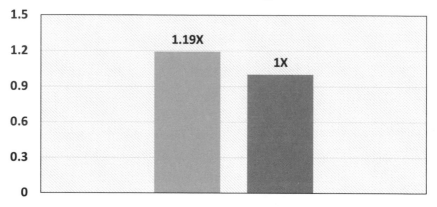

Relative Risk of Chorioamnionitis in Pregnant Women Vaccinated With Tdap Versus Unvaccinated Pregnant Women

■ Received Tdap Vaccine   ■ No Tdap Vaccine

Figure 11.9—Relative risk of chorioamnionitis in pregnant women vaccinated with the Tdap vaccine versus unvaccinated pregnant women (Kharbanda et al. 2014).

Figure 11.9 shows the results from the paper "Evaluation of the Association of Maternal Pertussis Vaccination with Obstetric Events and Birth Outcomes," published in the *Journal of the American Medical Association* in 2014.[57] The lead author is Dr. Elyse Kharbanda, affiliated with HealthPartners Institute for Education and Research in Minneapolis, Minnesota. The CDC funded the study. Kharbanda used the CDC's Vaccine Safety Datalink to evaluate pregnancy outcomes in women receiving the recommended maternal Tdap vaccine. Among women who received Tdap at any time during pregnancy, 6.1% experienced chorioamnionitis, compared to only 5.5% of women who were not inoculated with the Tdap vaccine.[58] When accounting for other vaccines in pregnancy, Tdap-vaccinated women had a statistically significant elevation in relative risk of chorioamnionitis at 1.19 and a 95% confidence interval of 1.13 to 1.26.[59] Chorioamnionitis, inflammation of the membrane that encapsulates a fetus in utero, is a dangerous condition predominantly associated with bacterial infections in either or both the mother and fetus that can result in preterm birth or stillbirth.[60]

Prenatal Tdap Immunization and Risk of Maternal and Newborn Adverse Events

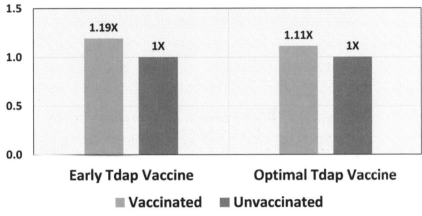

Figure 11.10—Risk ratios for chorioamnionitis in pregnant women receiving the Tdap vaccine optimally and early versus unvaccinated pregnant women (Layton et al. 2017).

Figure 11.10 shows results from the paper "Prenatal Tdap Immunization and Risk of Maternal and Newborn Adverse Events," published in *Vaccine* in 2017.[61] Dr. J. Bradley Layton, the lead author, is from the Department of Epidemiology at the University of North Carolina in Chapel Hill. In a large cohort study of pregnant US women, Layton found a statistically significant relationship between chorioamnionitis and optimally (at or after 27 weeks gestation age)

Tdap-vaccinated mothers with a risk ratio of 1.11 and a 95% confidence interval of 1.07 to 1.15 when compared to those who received no prenatal Tdap vaccine.[62] Those mothers receiving the Tdap vaccine early (before 27 weeks gestational age) showed a risk ratio of 1.19 and a 95% confidence interval of 1.11 to 1.28 when compared to those who received no prenatal Tdap vaccine.[63] The authors did not adjust their analysis to account for women receiving the influenza vaccine during pregnancy. Approximately 50% of women receiving the Tdap vaccine during pregnancy also received the influenza vaccine, whereas only 18% of women not receiving the Tdap vaccine during pregnancy received the influenza vaccine.

Figure 11.11 also shows results from the paper "Prenatal Tdap Immunization and Risk of Maternal and Newborn Adverse Events," published in the journal *Vaccine* in 2017.[64] Layton, the lead author, reported a greater incidence of post-partum hemorrhage associated

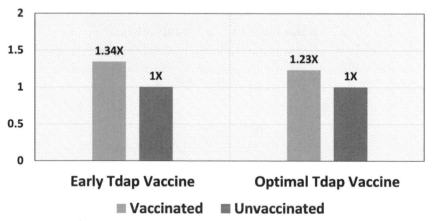

Figure 11.11—Risk ratios for post-partum hemorrhage in pregnant women receiving the Tdap vaccine early and optimally versus unvaccinated pregnant women (Layton et al. 2017).

with early Tdap vaccination (earlier than 27 weeks gestation), with a risk ratio of 1.34 and a 95% confidence interval of 1.25 to 1.44.[65] Those mothers receiving the Tdap optimally (after 27 weeks gestation) showed a risk ratio of 1.23 and a 95% confidence interval of 1.18 to 1.28 when compared to those who received no prenatal Tdap vaccine.[66] To give an idea of the magnitude of the type of problem identified in this study, this hazard ratio would result in an additional 29,000 cases of post-partum hemorrhage per year in the US if all pregnant women received the Tdap vaccine.

Maternal Tdap Vaccination and Risk of Infant Morbidity

Figure 11.12 shows results from the paper "Maternal Tdap

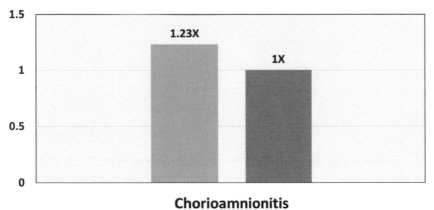

Figure 11.12—Hazard ratio for chorioamnionitis in pregnant women receiving the Tdap vaccine versus unvaccinated pregnant women (DeSilva et al. 2017).

Vaccination and Risk of Infant Morbidity," published in the journal *Vaccine* in 2017.[67] The lead author, Dr. Malini DeSilva, is affiliated with HealthPartners in Minneapolis, Minnesota. The CDC directly funded this study and provided data from the Vaccine Safety Datalink.

Within a cohort of nearly 200,000 pregnant women, DeSilva affirmed a greater rate of chorioamnionitis in pregnant women who received Tdap, with an adjusted rate ratio of 1.23 and a 95% confidence interval of 1.17 to 1.28.[68]

Enhanced Surveillance of Tetanus Toxoid, Reduced Diphtheria Toxoid, and Acellular Pertussis (Tdap) Vaccines in Pregnancy in the Vaccine Adverse Event Reporting System (VAERS), 2011–2015

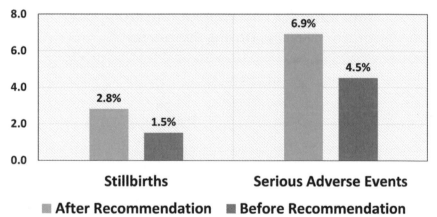

Figure 11.13—Percentage of stillbirths and serious adverse events reports on VAERS before and after the ACIP recommendation for the prenatal Tdap vaccine (Moro et al. 2016).

Figure 11.13 shows results from the paper "Enhanced Surveillance of Tetanus Toxoid, Reduced Diphtheria Toxoid, and Acellular Pertussis (Tdap) Vaccines in Pregnancy in the Vaccine Adverse Event Reporting System (VAERS), 2011–2015," published in the journal *Vaccine* in 2016.[69] The lead author, Dr. Pedro Moro, is an epidemiologist in the Immunization Safety Office of the CDC. Moro compared VAERS reports of adverse pregnancy effects before and after the CDC's Advisory Committee on Immunization Practices (ACIP) recommended administering the Tdap vaccine during the third trimester of pregnancy. The study authors observed an uptick in reports of stillbirths from 1.5% to 2.8% of all pregnancies, as well as an increase in serious adverse events from 4.5% to 6.9%.[70] Unfortunately, the study authors dismissed these findings "given the broader use of Tdap in pregnant women in the third trimester" and based on built-in limitations of VAERS, "underreporting, reporting biases and inconsistency in the quality of reports."[71]

Evaluation of Acute Adverse Events after COVID-19 Vaccination during Pregnancy

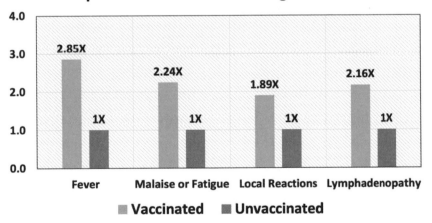

Rate Ratios of Adverse Events following COVID-19 Vaccination in Pregnant Women Compared to Unvaccinated Pregnant Women

Figure 11.14—Local and systemic reactions following COVID-19 vaccination in pregnant women compared to unvaccinated pregnant women (DeSilva et al. 2022).

Figure 11.14 shows results from the paper "Evaluation of Acute Adverse Events after COVID-19 Vaccination during Pregnancy," published in the *New England Journal of Medicine* in 2022.[72] Dr.

Malini DeSilva is the lead author from HealthPartners Institute in Bloomington, Minnesota. The CDC financially supported this research. Clinical trials did not specifically test COVID-19 vaccines used in the United States on pregnant women. The Comirnaty vaccine package insert makes this very clear.[73] However, the CDC recommends COVID-19 vaccination "for people who are pregnant, breastfeeding, trying to get pregnant now, or might be pregnant in the future."[74] Pregnant women receiving the COVID-19 vaccine, compared to matched, unvaccinated pregnant women, were 2.85 times more likely to experience fever (95% CI of 1.76 to 4.61), 2.24 times more likely to experience malaise or fatigue (95% CI of 1.71 to 2.93), 1.89 times more likely to sustain local reactions (95% CI of 1.33 to 2.68), and 2.16 times more likely to experience lymphadenopathy (swollen lymph nodes) (95% CI of 1.42 to 3.28).[75] The study's authors followed the cohort for 42 days after vaccination, precluding the evaluation of long-term adverse events.

Figure 11.15 shows results from the paper "Safety of Third SARS-CoV-2 Vaccine (booster dose) during Pregnancy," published in the *American Journal of Obstetrics and Gynecology* in 2022.[76] Lead author, Dr. Aharon Dick, is from the Department of Obstetrics and Gynecology, Hadassah Medical Organization and Faculty of Medicine, Hebrew University of Jerusalem in Israel. The researchers investigated 5,618 pregnant women, 2,305 who were vaccinated and 3,313 who were unvaccinated. Within the study, pregnant women fully vaccinated and boosted (i.e., triple vaccinated) with either the Pfizer BNT162b2 or the Moderna mRNA-1273 COVID-19 vaccine were three times as likely to experience post-partum hemorrhage (heavy bleeding after giving birth) as unvaccinated pregnant women (p-value < 0.001).[77] In addition, medical practitioners diagnosed triple-vaccinated pregnant women with gestational diabetes (high blood sugar) 1.5 times more often than unvaccinated pregnant women (p-value = 0.02).[78] Gestational diabetes can increase the risk

Safety of Third SARS-CoV-2 Vaccine (booster dose) during Pregnancy

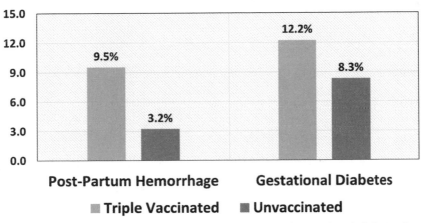

Rates of Post-Partum Hemorrhage and Gestational Diabetes Associated with COVID-19 Vaccination During Pregnancy

Figure 11.15—Rates of post-partum hemorrhage and gestational diabetes in COVID-19 triple-vaccinated and unvaccinated pregnant women (Dick et al. 2022).

of high blood pressure during pregnancy and lead to labor and delivery complications as well as preterm delivery.[79]

Figure 11.16 shows the results of an unpublished analysis of the CDC's Vaccine Adverse Event Reporting System.[80] Since introducing the first COVID-19 vaccine in December 2020, practitioners and patients have made 3,576 reports of spontaneous abortion due to COVID-19 shots.[81] This starkly contrasts with 1,089 reports of spontaneous abortion for all other vaccines over the 32-year history of VAERS. In addition, individuals have also submitted 19,040 reports of fertility disorders after receiving the COVID-19 vaccine versus 1,423 such reports for all other vaccines over the 32-year history of VAERS.[82] We completed these analyses using VAERS reports updated as of April 7, 2023.

VAERS Analysis of COVID-19 Vaccination in Pregnant Women

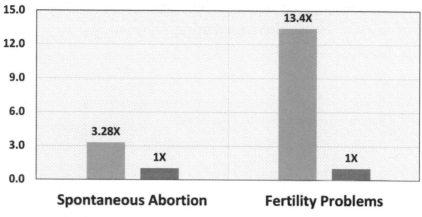

Figure 11.16—Ratio of all VAERS reports for COVID-19 vaccination versus all VAERS reports for all other vaccines combined over the 32-year history of VAERS (as of April 7, 2023).

Figure 11.17 shows results from the paper "COVID-19 Vaccination BNT162b2 Temporarily Impairs Semen Concentration and Total Motile Count among Semen Donors," published in the journal *Andrology* in 2022.[83] The lead author, Dr. Itai Gat, is affiliated with the Sperm Bank and Andrology Unit at the Shamir Medical Center in Tzrifin, Israel. Sperm concentration in semen of male donors reduced by 15.4% from before vaccination to 75–125 days after vaccination (p-value = 0.01 and a 95% CI of -25.5 to 3.9%).[84] In addition, total motile sperm count decreased by 22.1% over the same duration (p-value = 0.007 and a 95% CI of -35.0 to 6.6).[85] These reductions were statistically significant. Both sperm concentration and total motile sperm count remained at a reduced rate after 150 days, with decreases of 15.9% and 19.4%, respectively.[86] However, these results were not statistically significant due to high variability in measurement and fewer subjects providing samples.

COVID-19 Vaccination BNT162b2 Temporarily Impairs Semen Concentration and Total Motile Count among Semen Donors

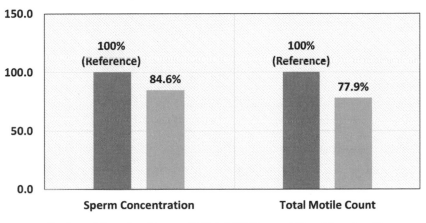

Figure 11.17—Ratio of sperm concentration and total motile sperm count before COVID-19 vaccination and 75–125 days after COVID-19 vaccination (Gat et al. 2022).

Because of this reduced statistical power, the authors' assertion that recovery of semen parameters was evident after 150 days is also not supported.

Summary

Fetal loss is associated with influenza vaccination in specific instances in four of the papers cited in this chapter. Irving found an association when practitioners administered the flu vaccine prior to conception.[87] Goldman[88] and Donahue[89] observed associations with the H1N1 vaccine, and Donzelli[90] reported associations with the trivalent inactivated vaccine. Zerbo linked influenza vaccination to autism spectrum disorder in offspring.[91] Price also observed such a relationship when looking at thimerosal exposure through the influenza vaccine and anti-RhoD

	Zerbo et al. 2017	Irving et al. 2013	Goldman 2013	Donahue et al. 2017	Donzelli et al. 2019a	Donzelli et al. 2019b	Christian et al. 2011	Price et al. 2010	Layton et al. 2017
ASD	✓							✓	
Fetal Loss		✓	✓	✓		✓			
Serious Adverse Event					✓				
Inflammatory Markers							✓		
Chorioamnionitis									✓

Table 11.1 - Summary of results comparing health outcomes of pregnant women exposed to the influenza vaccine. Significantly higher odds ratios, relative risks, hazard ratios, or incidences are denoted by a ✓.

	Kharbanda et al. 2014	Layton et al. 2017	DeSilva et al. 2017	Moro et al. 2016	DeSilva et al. 2022	Dick et al. 2022	VAERS 2022	Gat et al. 2022
Chorioamnionitis	✓	✓	✓					
Post-Partum Hemorrhage		✓					✓	
Fetal Loss				✓			✓	
Serious Adverse Event				✓				
Fever					✓			
Malaise					✓			
Local Reaction					✓			
Lymphadenopathy					✓			
Gestational Diabetes							✓	
Fertility Problems							✓	✓

Table 11.2 - Summary of results in comparing health outcomes of pregnant women exposed to the Tdap vaccine (white) or COVID-19 vaccine (yellow). Significantly higher odds ratios, relative risks, hazard ratios, or incidences are denoted by a ✓.

immunoglobulin given during pregnancy.[92] Researchers reported serious adverse events,[93] increased inflammatory markers (C-reactive protein and TNF-α),[94] and chorioamnionitis in each highlighted study. [95]

Three studies reported that pregnant women exposed to the Tdap vaccine had higher incidences of chorioamnionitis compared to unvaccinated pregnant women.[96, 97, 98] Layton[99] reported a higher incidence of post-partum hemorrhage, and Moro[100] reported upticks in VAERS reports of fetal loss and serious adverse events following the ACIP approval of the Tdap vaccine for pregnant women. COVID-19 vaccination was associated with fetal loss,[101] post-partum hemorrhage,[102] gestational diabetes,[103] and fertility problems.[104, 105] Gat reported lower sperm count in men following COVID-19 vaccination.[106] DeSilva did not report long-term sequelae to COVID-19 vaccination but did report increases in short-term COVID-19 vaccine reactions, including fever, malaise, local reaction, and lymphadenopathy.[107]

Afterword by Children's Health Defense Staff

Vaccine Safety Project—Six Steps Overview

"Autism, ADHD, epilepsy, autoimmune disorders, deadly allergies, SIDS, juvenile rheumatoid arthritis, diabetes, learning disabilities and more have been increasing for over thirty years. Over 50 percent of our children are chronically ill. An NIH study found that 49.5 percent of adolescents ages thirteen to eighteen have a mental disorder. This is unacceptable."

—Robert F. Kennedy Jr.

The **Vaccine Safety Project** of Children's Health Defense is an outcome of an investigative review of the US government's vaccine approval/recommendations process and post-marketing safety monitoring. CHD and Robert F. Kennedy Jr. formulated these six steps as necessary recommendations for improving vaccine safety and protecting children from vaccine injuries. We hope that they will be used by parents and vaccine safety advocates as tools to educate local policy makers, state and federal legislators, and public health officials who need to know the facts about our federal vaccine programs, vaccine safety, and the Vaccine Injury Compensation Program.

The long-term health effects of our vaccine program are inadequately studied, and our regulatory bodies are conflicted. Childhood health epidemics have mushroomed along with the childhood vaccine schedule. Vaccines contain many ingredients, some of which are known to be neurotoxic, carcinogenic, and cause autoimmunity.

Vaccine injuries can and do happen. As of June 1, 2023, the National Vaccine Injury Compensation Program of Health and Human Services (HHS) has awarded approximately $5 billion for vaccine injuries since 1988.[1]

"This is a call to action to all vaccine safety advocates to effect lasting change in global public health policy. Our efforts now will protect the lives of future generations."

—Brian S. Hooker, PhD

6 Steps to Vaccine Safety

1	Subject vaccines to the same rigorous approval process as other drugs.
2	Mandatory reporting of vaccine adverse events and automate the VAERS* and VSD* databases.
3	Ensure everyone involved with federal vaccine approvals and recommendations are free from conflicts of interest.
4	Reevaluate all vaccines recommended by the ACIP* prior to the adoption of evidence-based guidelines.
5	Study what makes some individuals more susceptible to vaccine injury.
6	Support fully informed consent and individual rights to refuse vaccination.

*VAERS: Vaccine Adverse Events Reporting System, ***VSD**: Vaccine Safety Datalink,
*ACIP: Advisory Committee on Immunization Practices

Common sense dictates that these Six Steps to Vaccine Safety must be taken:

1. Subject vaccines to a scientifically rigorous approval process.

Vaccines are regulated by the FDA's Center for Biologics Evaluation and Research (CBER) division as "biologics"[2] and are not always put through the same level of safety testing as new pharmaceuticals, which are regulated under the Center for Drug Evaluation and Research (CDER).[3, 4]

- Vaccines, which are given to healthy patients, should be tested more rigorously than drugs because they are not given to treat an existing disease.
- Inadequate testing currently ensures that the true risk/benefit assessments for the safety and cost of vaccines are impossible to calculate accurately.

These vaccines are given to about 4 million American infants annually.

Typical Drug Approval Process	Typical Vaccine Approval Process
Prelicensure follow-up for adverse events often takes years. For example: Lipitor—4.8 years[5] Enbrel—6.6 years[6] Stelara - 5 years[7]	Prelicensure follow-up for adverse events may take as little as 2–5 days. For example: HepB (Engerix—GSK)—**4 days**[8] HepB (Recombivax—Merck)—**5 days**[9] Polio (IPOL—Sanofi Pasteur)—**2 days**[10] Hib (Pedvax—Merck)—**3 days**[11] Hib (Hiberix—GSK)—**4 days**[12] Hib (ActHib—Sanofi Pasteur)—**30 days**[13]
Requirement for trials to be done against an inactive placebo—with the exception of drugs for life-threatening diseases (cancer, etc.) where the placebo is typically the current standard of care.	Trials not done against an inactive placebo. Trials of vaccinated compared to unvaccinated children are not performed.

(Continued on next page)

Typical Drug Approval Process	Typical Vaccine Approval Process
Placebo is often: • Saline • A sugar pill designed to look like the active pill • Another inactive substance or base	Placebo is often: • Another vaccine, but not always for the same disease • An adjuvant or preservative like aluminum or mercury that is not inactive • A group of vaccines
Safety follow-up is incentivized by education and lawsuits. There are free market checks and balances to produce safer drugs.	Lack of any product liability for vaccine manufacturers provided by the National Childhood Vaccine Injury Act **eliminates market incentives to produce safe vaccines.**

2. Require reporting of vaccine adverse events. Automate VAERS and VSD databases for research.

Reporting and study of adverse events after receipt of vaccines is currently haphazard and antiquated. Since these two databases are the primary sources of US post-licensure surveillance, serious side effects of vaccination that were unclear or not seen in clinical trials will be missed.

The Vaccine Adverse Events Reporting System (VAERS) is the online system into which doctors and patients report adverse events after vaccination. HHS admits that the system *likely records only about 1% of the actual adverse events,* but even after a three-year HHS/AHRQ study showed the feasibility of automating reports using electronic medical records,[14] Centers for Disease Control (CDC) has been nonresponsive to "multiple requests to proceed with testing and evaluation."

- Clinical trials for vaccines typically only enroll a few thousand patients in total. When vaccines are subsequently approved for use in populations of millions of healthy individuals, it is imperative that rates of known adverse events and any new or rare adverse events be monitored.
- Without adequate safety follow-up, serious side effects may be missed entirely, putting the public at risk (examples of the past

importance of safety follow-up include hormone replacement therapy, Vioxx, and amphetamines).

- There has never been a comparative study of broad health outcomes in vaccinated vs. unvaccinated populations.

The National Childhood Vaccine Injury Act (NCVIA) requires health-care providers to report:[15]

- Any adverse event listed by the vaccine manufacturer as a contraindication to further doses of the vaccine; or
- Any adverse event listed in the VAERS Table of Reportable Events following Vaccination that occurs within the specified time period after vaccination.

But, in practice, this doesn't happen. There is no consequence for failing to report an injury. There is no mechanism for prosecution of non-compliance and, therefore, no incentive for a busy doctor to report vaccine safety problems.

The Vaccine Safety Datalink (VSD) is a collaborative project between CDC's Immunization Safety Office and eight private health-care organizations. The VSD was started in 1990 to monitor the safety of vaccines and conduct studies about rare and serious adverse events following immunization.[16] However, research is currently hampered by the lack of broad access to this publicly funded database, the variability of reporting, and the statistical structure of the database.

3. Ensure all parties involved with federal vaccine approvals and recommendations are free from conflicts of interest.
FDA's Vaccine and Related Biological Products Advisory Committee (VRBPAC) is responsible for licensing of vaccines. CDC's Advisory

Committee on Immunization Practices (ACIP) is responsible for adding vaccines to the recommended schedules.

- CDC or NIH Employees whose names appear on vaccine patents can receive *up to $150k in licensing fees* per year (in perpetuity).[17]
- Regarding VRBPAC, a House OGR Committee Report found that the *"overwhelming majority of members, both voting members and consultants have [sic] substantial ties to the pharmaceutical industry,"* and "committee members with substantial ties to pharmaceutical companies have been given waivers to participate in committee proceedings."[18]
- A similar report on the ACIP found that *"The CDC grants blanket waivers to the ACIP members each year that allow them to deliberate on any subject, regardless of their conflicts, for the entire year."*[19]
- A 2009 HHS Office of the Inspector General report found that:[20]
  –"CDC had a systemic lack of oversight of the ethics program."
  –97% of committee members' conflict disclosures had omissions.
  –58% had at least one unidentified potential conflict.
  –32% had at least one conflict that remained unresolved.
  –CDC continued to grant broad waivers to members with conflicts.

All vaccine regulatory agencies must rigorously enforce their ethics policies to ensure that our vaccine program is free from financial conflicts of interest.

4. Reevaluate all vaccines recommended by ACIP prior to the adoption of evidence-based guidelines.

A vote by the Advisory Committee on Immunization Practices results in:

- Mandating the vaccine to millions of children
- Immunity from liability for the manufacturers
- Inclusion in the Vaccines for Children program

However, prior to 2012, ACIP did not use evidence-based guidelines to evaluate their vaccine recommendations. Evidence Based Practice is "the conscientious, explicit and judicious use of current best evidence in making decisions about the care of the individual patient. It means integrating individual clinical expertise with the best available external clinical evidence from systematic research."[21] The final ACIP guidelines published in November of 2013 outlined clearly, for the first time, a standardized plan to evaluate the quality and strength of the research behind each recommendation for a vaccine for each population. ACIP's recommendations include the populations, timing, spacing, number of doses, boosters, and appropriate ages for each vaccine to be administered.

The CDC's infant schedule, given to approximately 4 million babies a year, was largely adopted before these guidelines were in place. Vaccines recommended before the adoption of evidence-based guidelines should not have been "grandfathered" in. Earlier ACIP recommendations should be thoroughly reviewed in light of the new guidelines and current research.

5. Study what makes some individuals more susceptible to vaccine injury.

The Institute of Medicine (now the National Academy of Medicine) has issued three disturbing reports, in 1991, 1993, and 2011, on the evidence for suspected and/or reported vaccine adverse events:

Year	Vaccine(s) Studied	# of Conditions Reviewed	Evidence Supports Vaccine Causation	Evidence Supports Rejection of Vaccine Causation	Evidence Inadequate to Accept or Reject Vaccine Causation
1991[22]	DPT, MMR	22	6	4	12
1993[23]	DT, MM, MMR, HepB, Hib	54	12	4	38
2011[24]	Varicella, T, HepB, MMR	155	16	5	134
Totals		231	34	13	184

- In 2013, the IOM studied the entire Childhood Immunization schedule and stated:

  "No studies have compared the differences in health outcomes . . . between entirely unimmunized populations of children and fully immunized children. . . . Furthermore, studies designed to examine the long-term effects of the cumulative number of vaccines or other aspects of the immunization schedule have not been conducted."[25]

- The Vaccine Injury Compensation Program has paid out approximately $5 billion in compensation to victims of vaccine injury. The children and adults who have been compensated for injuries have never been studied to determine why they were injured, in an effort to make vaccines safer for everyone. Preventing vaccine injuries should be tackled as zealously as we tackle preventing infectious diseases.

Vaccine safety science, particularly long-term safety science, is inadequate to ensure children's safety or to accurately assess risks for purposes of informed consent.

6. Support fully informed consent and individual rights to refuse vaccination.

The American Academy of Pediatrics' statement on the ethics of informed consent includes the following stipulation: "patients should have explanations, in understandable language, of . . . ; *the existence and nature of the risks involved;* and the existence, potential benefits, and risks of recommended alternative treatments (*including the choice of no treatment*)."[26]

- In the case of vaccination, informed consent is often ignored completely in real-world settings.

  By law, "*all health care providers in the United States who administer, to any child or adult, any of the following vaccines—diphtheria, tetanus, pertussis, measles, mumps, rubella, polio, hepatitis A, hepatitis B, Haemophilus influenzae type b (Hib), influenza, pneumococcal conjugate, meningococcal, rotavirus, human papillomavirus (HPV), or varicella (chickenpox)—shall, **prior to administration** of each dose of the vaccine, provide a copy to keep of the relevant current edition vaccine information materials that have been produced by the Centers for Disease Control and Prevention (CDC) to the parent or legal representative1 of any child to whom the provider intends to administer such vaccine, or to any adult to whom the provider intends to administer such vaccine.*"[27]

- In practice, particularly when multiple vaccines are administered on the same day, many parents report that they got the Vaccine Information Sheet (VIS) as they left and there was no explanation of information before a vaccine was given. It is also rare that medical history is thoroughly discussed to identify contraindications to a vaccine. For example, a patient with a family history of autoimmunity is likely at increased risk for an autoimmune reaction after vaccination.

The following are examples of the types of information that patients may learn after the fact from the Vaccine Information Sheets:

- "Severe events have very rarely been reported following MMR vaccination, and might also happen after MMRV. These include: *Deafness, long-term seizures, coma, lowered consciousness, brain damage.*"
- Or this from the Polio VIS and several others: "As with any medicine, there is a very remote chance of a vaccine causing a serious injury or death."

Lack of informed consent encompasses vaccine advertising, as well. While television drug ads disclose the side-effect risks of that drug at length, vaccine advertising does not. The patient, again, is at a disadvantage.

Conclusion

Insistence on fully informed consent and individual rights to refuse a vaccination become imperative given the lack of long-term follow-up and surveillance, only 1% of adverse events are captured and reported,[28] vaccine recommendations are tainted by financial conflicts of interest of regulators, the current childhood vaccine schedule was not approved using evidence-based science and policy, the childhood vaccine schedule has never been tested on fully vaccinated vs. unvaccinated, and there is sparse research into which patients are likely to have adverse events. America is in the midst of many childhood epidemics. Over 50% of our children are chronically ill.[29] We owe it to our children to examine what is happening to their health and correct it as soon as possible.

APPENDIX A

Missed Opportunities: Aftermath of the May 2017 NIH Meeting with Collins, Fauci, et al.

As word spread that President-elect Donald Trump had invited Robert F. Kennedy Jr. in January 2017 to lead Trump's proposed Vaccine Safety Commission, there were two distinct reactions from around the nation: hope among vaccine safety advocates and parents of vaccine injured children; and outrage from mainstream medicine, public health officials, and others who benefit from the pharmaceutical industry's billowing bottom line. More than 350 medical groups including the American Academy of Pediatrics wrote a letter to Trump on February 7 insisting that vaccines are safe and that instead of investigating them, the nation should "redouble our efforts to make needed investments in patient and family education about the importance of vaccines in order to increase the rate of vaccination among all populations."[1] Pharma friend-in-chief Bill Gates later bragged that in March of 2017, he had advised President Trump that establishing a Vaccine Safety Commission would be ". . . a dead end. That would be a bad thing. Don't do that."[2]

Following the May 2017 meeting with the NIH, despite know-
ing that the deck was stacked against them, Kennedy and Children's
Health Defense pressed on with their efforts to educate Dr. Collins and
Dr. Fauci on the need to conduct vax-unvax studies and to undertake
more rigorous vaccine safety research. Kennedy presented the rationale and scientific justification for such actions clearly in the follow-up
correspondence reproduced below along with the sole, anemic response
from the NIH. Today, the often-devastating and ongoing impacts
from our nation's response to the COVID crisis, including the relentless bombardment of government and industry propaganda regarding
the "safety and efficacy" of COVID shots, beg the question: could the
suffering among American citizens have been avoided or ameliorated
had the Vaccine Safety Commission been allowed to go forward?

Email from Robert F. Kennedy Jr. to Dr. Francis Collins, NIH Director, 6/21/17[1]

From: Robert Kennedy Jr.
Date: Wed, Jun 21, 2017 at 8:50 PM
Subject: Re: Follow up on vaccine data accessibility
To: Collins, Francis (NIH/OD) [E]

Dear Francis,

Your email below is emblematic of the core issue we raised during our meeting regarding the expanding gaps in vaccine safety science and HHS's refusal to fill those gaps. During our two-hour meeting with White House and senior HHS officials, we laid out some of the significant deficiencies in safety testing and surveillance (both pre and post-licensure) and the well-documented conflicts of interest at CDC and FDA that prevent those agencies from addressing these issues, including their refusal to mandate the kind of fundamental safety studies it requires of every other pharmaceutical product. True to that concern, rather than proposing solutions to address even a single of these deficits, your response attempts to justify the assertion that basic vaccine safety science simply *cannot* be performed. This is particularly troubling as you are the head of the Federal Government Task Force responsible for recommending to the Secretary of HHS ways to improve vaccine safety, including safety testing and monitoring.

In any event, your claims below regarding the Vaccine Safety Datalink ("**VSD**") are incorrect and give rise to a number of important issues I now address in turn:

1. VSD Vaccinated vs. Unvaccinated Study:

- Your claim that the Institute of Medicine ("**IOM**") has found that the VSD database cannot be used to study vaccinated versus unvaccinated children is simply untrue. In fact, IOM specifically concluded, in the very report you cite, that such a study was possible: "some stakeholders have suggested that further research is warranted, such as a comparison of vaccinated children with unvaccinated children or children immunized on alternative schedules. It is possible to make this comparison through analyses of patient information contained in large databases such as VSD."[2] That report even provides that: "The most feasible approach to studying the safety of the childhood immunization schedule is through analyses of data obtained by VSD."[3] Your irrelevant quote from the same IOM report merely concerns the IOM's analysis of a potential prospective cohort study (i.e., randomized controlled trial) of small isolated populations. The quote you cited has no bearing on our proposed retrospective study utilizing the VSD.

- Your next claim that there are an insufficient number of unvaccinated children in the VSD is also untrue. A CDC study from 2013 found that of children in the VSD born between 2004 and 2008, approximately 50% were under-vaccinated (around 160,000 children) and that approximately 1% of children (around 3,200 children) were completely unvaccinated in the first two years of life. This 2013 study was limited to only a four-year period, but the VSD has 25 years of data. Hence, the total number of under-vaccinated and unvaccinated children is far greater than that reported in the 2013 study and is certainly large enough to allow for a sufficient retrospective study.[4]

- Your third claim that outside researchers can access the VSD by meeting some basic requirements is also misleading and untrue. While we agree with the fundamental principle of limiting access to the data in ways that assure privacy and safety of the information, HHS has imposed restrictive criteria that makes access by independent scientists effectively impossible. The CDC has systematically blocked, delayed and undermined any outside researchers access to the VSD. Indeed, in the past 17 years, we know of only two researchers independent of the CDC that have received access to the VSD, and only after persistent congressional intervention, and their treatment by the CDC thereafter was found in a congressional letter to be "abysmal and embarrassing." (See attached document describing some of their ordeals.) Furthermore, to the extent that access to the VSD is allowed, the CDC website states that only "VSD data created before 2000 are available through the data sharing program for new vaccine safety studies for analyses at the RDC."[5] This arbitrary limitation makes it impossible to conduct a valid longitudinal vaccine safety study. There is no logical reason why the HHS refuses to grant access to the last 17 years' worth of data. (In contrast to the restrictions placed on outside researchers, the CDC has published hundreds of papers to exonerate certain vaccines and certain vaccine ingredients of claims they cause harm, using VSD data and then, contrary to all scientific protocols, refuses to release the raw data underlying these studies when they are challenged.)

- Finally, if your argument that confounding variables make it impossible to conduct studies of children with different vaccination status in the VSD is correct, then a significant portion of the CDC's over 200 studies using the VSD are invalid, including every single one of the CDC's thimerosal and MMR studies.[6]

The purpose of the VSD is to provide a repository of information to assess vaccine safety. American taxpayers pay a reported $27 million annually to maintain the VSD for this purpose. *Are you really now claiming that the archival methodologies employed by the agency under your leadership have rendered the VSD inadequate to that purpose? Are you also really arguing that the brain trust at NIH cannot devise a way to take the same safety precautions for vaccines that are required for every single drug: compare the actual health outcomes of those receiving vaccines with those that have not received vaccines?*

2. NIH Director's Statutory Obligation to Recommend Improvements to Vaccine Safety.

As I am sure you are aware, the United States Code expressly creates a Task Force on Safer Childhood Vaccines (the "**Task Force**") of which you, Francis, are the Chairman:

The Secretary shall establish a task force on safer childhood vaccines which shall consist of the Director of the NIH, the Commissioner of the FDA, and the Director of the CDC. The Director of the NIH shall serve as chairman of the task force . . . [and] shall prepare recommendations to the Secretary [of HHS] concerning implementation of the requirements of subsection (a) . . . [to] promote the development of childhood vaccines that result in fewer and less serious adverse reactions . . . [Dec. 22, 1987] and promote the refinement of such vaccines, and (2) make or assure improvements in, and otherwise use the authorities of the Secretary with respect to, the licensing, manufacturing, processing, testing, labeling, warning, use instructions, distribution, storage, administration, field surveillance, adverse reaction reporting, . . . and research on vaccines, in order to reduce the risks of adverse reactions to vaccines.

(42 U.S.C. § 300aa-27.) This law clearly makes you the person responsible for developing recommendations, in all manner and form, to the HHS Secretary to improve vaccine safety. Your follow-up below seeking to avoid the most basic study of vaccine safety raises questions regarding your commitment to the foregoing responsibilities. *Thus, we ask that you provide written copies of all recommendations to improve vaccine safety you have made to the Secretary of HHS while serving as the Director of NIH.*

3. Pre-Licensure Safety Testing.

During our meeting, we discussed several issues regarding pre-licensure safety testing for vaccines and await the additional support/explanation you and your team claimed existed. During our meeting you promised to provide us copies of these items.

a. Safety Data for Hepatitis B Vaccine Given to 1 Day Olds: During the meeting, Dr. Fauci intimated that there was pre-licensure safety testing of the two Hepatitis B vaccines given to one day old babies in the United States (Engerix and Recombivax) beyond the four and five day safety reviews disclosed by their manufacturers. If Dr. Fauci's statement is true, it's hard to understand why the manufacturers would withhold this information from the public on their package inserts and websites. Nonetheless, we accept Dr. Faurci at his word, yet we still have not received from you the purported additional pre-licensure safety data for Engerix and Recombivax. *When do you intend to provide that information?*

b. HPV Licensing Study Saline Placebo Data. As discussed during our meeting, the pre-licensure clinical trial for the HPV vaccine had a subject group (that received the HPV vaccine) and two control groups, one which received an injection of aluminum adjuvant (implicated in all forms of systemic autoimmune disorders) and another group that received a saline placebo. The differences between all three groups were reported for local reactions but for systemic autoimmune disorders, the data for the aluminum adjuvant and saline placebo groups were combined, hence concealing the actual systemic autoimmune adverse vaccine reaction rates between the HPV vaccine and the saline placebo groups. You stated with confidence that breaking down the systemic autoimmune disorder rate between these two control groups would not show a difference. As we answered at that time, we prefer to rely on the data rather than opinion or assumption, and hence would like to see the data. You indicated that you would provide it. We still have not received this information from you. To reiterate, we want to see the pre-licensure clinical trial data for the HPV vaccine, including all data reflecting the rate of systemic autoimmune disorders between the subject, aluminum-adjuvant-control and saline-control groups. *When do you expect to provide us with this data?*

c. Lack of Saline Placebo in Vaccine Clinical Trials. You vigorously defended the inadequate and scientifically invalid use of an aluminum adjuvant or another vaccine as a placebo in pre-licensure vaccine clinical trials. When we pointed out that the use of a potent neurotoxin or another vaccine instead of a true inert placebo in the control group would conceal dangerous side effects caused by the vaccine, you replied that you considered it "a brilliant design." I respectfully find that statement astounding. At best, using a spiked-aluminum-adjuvant or another vaccine as a placebo instead of an inert-saline-placebo (used in the clinical trial of every other drug) violates several standards of basic scientific protocols for drug testing. *Please explain how the actual safety profile of a vaccine can be determined from a clinical trial that does not use a saline placebo control?*

4. 134 Serious Common Adverse Reactions Reported Following Vaccination.

As discussed during the presentation, the IOM in 2011 reviewed the 155 most common serious adverse reactions complained of in the Vaccine Court and concluded that for 134 of these the science had not been done by HHS to confirm whether they are causally related to vaccination. The HHS is statutorily required to conduct such science. What steps have you taken or are you taking to confirm whether or not these 134 serious, and often devastating, conditions are causally related to vaccination?

5. CDC Refusal to Cooperate to Automate VAERS Reporting.

We brought to your attention during the meeting that, in 2010, the CDC refused to cooperate with a program to create a system to improve and automate reports submitted to the Vaccine Adverse Events Reporting System ("**VAERS**")—CDC's flawed passive/voluntary system for reporting vaccine injuries. HHS acknowledges that VAERS currently captures fewer than 1% of vaccine injuries. We showed you that another HHS agency spent nearly $1 million to create a pilot system that would automatically generate VAERS reports from electronic hospital records. The consulting group successfully implemented this pilot system at Harvard Pilgrim Health Care. After this pilot system was proven to work, and showed a shocking vaccine injury rate of nearly 10%, CDC cut off all contact with its designers and shut down the program. You indicated you could provide an explanation for the CDC's actions. *When do you intend to provide that explanation?*

6. Thimerosal.

During the month leading up to our meeting, we agreed to exchange published studies supporting our divergent positions on thimerosal safety. In accordance with that agreement, we provided you with 189 studies and reviews linking thimerosal to a range of neurodevelopmental and chronic diseases that have become epidemic in American children since the dramatic expansion of the vaccine program in 1989. We also gave you an additional 89 studies and reviews specifically linking thimerosal to autism. In turn, you provided us a list of random vaccine safety studies created by an industry advocacy group. Almost all these studies were pertinent to the MMR vaccine or other vaccines that do not contain thimerosal. Despite our repeated requests, you have not been able to produce a single study that demonstrates thimerosal safety. When I pressed you on this issue during our meeting you directed this question to the entire upper level NIH staff. NIH top officials and scientists were able to reference only a single study, Mady Hornig's study.[7] However, as I pointed out at that time, Hornig showed that she could induce autism like behaviors in mice by injecting them with thimerosal. That study, therefore, hardly supports your position. I once again renew my request that you provide us the studies that FDA relies upon to justify the continued unnecessary injection of this known neurotoxin into babies and pregnant women.

7. Autism:
a. Claim Vaccines Do Not Cause Autism. As with most vaccines (other than MMR) there has not been a single study regarding whether

DTaP causes autism. For example, the IOM in a 2011 report stated that the IOM could not confirm whether DTaP causes autism because no science had been done on that point. Nevertheless, the HHS baldly claims that all "Vaccines Do Not Cause Autism."[8] Therefore, can you please explain how HHS claims that vaccines do not cause autism when it does not know whether DTaP causes autism?

b. Genetics Research Re: Autism at NIH: The NIH's focus regarding autism concerns efforts to find genetic causes and not environmental causes for the condition. If autism was primarily the result of genetics, rather than some change in the environment, one would expect that the autism rate would have remained relatively steady over the centuries. *Is it the position of the NIH that autism has remained relatively steady over the past few centuries?*

8. Reducing Conflicts of Interest in Vaccine Policy:

During our meeting we reviewed government reports documenting conflicts of interest in vaccine safety at HHS and hence requested support for the following: (a) prohibiting any conflict-of-interest waivers for members of HHS's vaccine committees (ACIP, VRBPAC, NVAC and ACCV), (b) requiring members of HHS's vaccine committees to contractually agree not to accept any compensation, directly or indirectly, from any vaccine manufacturer for at least five years, and/or (c) require that vaccine-safety-advocates comprise at least 50 of HHS's vaccine committees. *Please inform us whether you support these options, or any other options, for limiting or eliminating the conflicts of interest on the HHS vaccine committees?*

Our overall ask is very simple and we think should not be controversial:

- eliminate conflicts at HHS regarding vaccine safety
- conduct proper pre-licensure safety studies

- conduct proper post-licensure safety surveillance
- conduct studies of the vaccine schedule to determine whether or not it is contributing, and to what degree, to the massive upsurge in childhood immunological and neurological disorders.

Your follow-up email below to the foregoing requests simply stated that HHS can't test vaccines retrospectively, and can't monitor them prospectively. Given the inadequate pre-licensure safety testing that is conducted and lack of post-licensure surveillance, your response presents the inevitable conclusion that families just have to accept vaccines developed and administered in a scientific cloak of darkness. What will America think when they learn that the "leaders" of medical science in this country cannot design a study to understand how vaccines may be affecting their overall health.

It is apparent that you have, at best, the most meager and anemic scientific basis for your assertion that vaccines are not contributing to the upsurge in childhood immunological and neurological disorders currently impacting over 50% of American children. Given the gravity of this issue, shouldn't we undertake the more robust science the IOM and other serious organizations has doggedly requested of you and other HHS agencies?

The concern in this country over vaccine safety and distrust of the HHS regarding vaccines is growing. That trend is likely to continue until HHS increases its transparency in this arena. The first step in restoring integrity and credibility to the vaccine program would be to provide thoughtful answers to the foregoing questions along with the basic documents and data regarding vaccine safety that we have requested. We look forward to receiving your timely reply.

Truly yours,
Robert F. Kennedy Jr.

Letter from Robert F. Kennedy Jr. to Dr. Francis Collins, NIH Director, 7/3/17[1]

United States Department of Health & Human Services
National Institutes of Health
Dr. Francis S. Collins, M.D., Ph.D.
Director of the National Institutes of Health
9000 Rockville Pike
Bethesda, Maryland 20892

July 3, 2017

Re: Vaccine Safety Meeting

Dear Dr. Collins,

First, I want to thank you for taking the time to meet with me and other representatives from the Children's Health Defense on May 31st.

I very much appreciate your efforts to pull together the meeting and your willingness to hear our concerns regarding vaccine safety.

My primary reason for writing is to address your staff's proposal for a longitudinal study to identify environmental toxins (including those in vaccines) that may be causing the childhood chronic disease epidemic. During our May 31st meeting, you and your staff suggested such a study as an alternative to our request that you open the Vaccine Safety Datalink (VSD) to independent scientists and to Verily, Google's medical research division, for machine analysis.

At your prodding, Dr. Linda Birnbaum from NIEHS and Dr. Diana Bianchi, Director of the *Eunice Kennedy Shriver* NICHD, proposed the alternative research project. Doctors Birnbaum and Bianchi contemplated a longitudinal prospective study that would follow mothers throughout their pregnancies and their infants in early life in an effort to identify risk factors that might result in adverse health outcomes, namely, the chronic disorders and neurodevelopmental disorders, including autism, now epidemic in nearly half of America's children.

We believe that such a study is a worthwhile endeavor. However, as we pointed out during the meeting, we are concerned that the initiative would require years to enroll mothers and collect the necessary data for analysis before reporting any findings. That relaxed time table does nothing to address the immediate health crisis. The CDC acknowledges that an astonishing one in six American children now suffers from a neurodevelopmental disorder[2] while an HHS funded study showed that 43% have a chronic disease, including allergy, diabetes and seizures.[3] These outbreaks—including the explosion of deadly food allergies that my own children suffer—suddenly became epidemic, coterminous with the dramatic expansion of the vaccine schedule. A study that will take at least a decade to provide answers does not seem like an acceptable solution.

Furthermore, many parents are frustrated that the NIH and other Federal agencies, such as the CDC, have repeatedly announced and commenced studies virtually identical to the one you now propose. Following great fanfare and loud launch announcements, these studies have largely underachieved. We fear that your latest proposal would be yet another redundant research initiative within the NIH or yet another deadend. The current and past studies include: CHARGE, MARBLES, EARLI, SEED, NCS and the latest, ECHO. Here is a short summary of each study:

The NIH launched The Childhood Autism Risks from Genetics and Environment (CHARGE) in 2003 to address a wide spectrum of chemical and biologic exposures and susceptibility factors, to assess environmental causes for autism, mental retardation and developmental delay.[4] The NIH touted CHARGE as the first ever major epidemiological case-control investigation to identify the culprits behind these disorders. The study, of up to 2,000 California children included detailed developmental assessments, medical information, questionnaire data, and collection of biologic specimens. Over 1,000 families were enrolled when the NIH discontinued funding for CHARGE in 2011. CHARGE researchers have published investigations into about 25 risk factors for autism but none of these studies has led to conclusive evidence or recommendations. CHARGE has omitted any studics of vaccines.

MARBLES (Markers of Autism Risk in Babies) was an extension of the CHARGE study launched in 2006.[5] The NIH provided a $7.5 million grant to fund this longitudinal study of pregnant women who had a biological child with autism spectrum disorder. As with your present proposal, the NIH announced that it meant to investigate pre-natal and post-partum biological and environmental exposures and risk factors that might contribute to the development of autism. MARBLES followed mothers before, during, and after their pregnancies, obtaining information about the pre-natal and post-natal environmental exposures. NIH researchers collected information about each participant's genetics and environment through blood, urine, hair, saliva, and breast milk, as well as home dust samples to obtain a comprehensive picture of the environment surrounding each pregnancy. The NIH also obtained information through interviews, questionnaires, and by accessing medical records. The study enrolled 450 mother-child pairs and then ended, without any noteworthy accomplishment, in 2011. The only published papers from MARBLES, 6 years after funding was discontinued, are two placental studies and an overview of the project. U.C. Davis has archived the specimens collected in both the CHARGE and MARBLES studies.

The NIH initiated the Early Autism Risk Longitudinal Investigation (EARLI) study with a $14 million Autism Centers of Excellence grant awarded by the National Institute of Environmental Health Sciences, the National Institute of Mental Health, the *Eunice Kennedy Shriver* National Institute of Child Health and Human Development, and the National Institute of Neurological Disorders and Stroke in 2008.[6] An additional $2.5 million grant came from Autism Speaks. Like MARBLES and CHARGE, the purpose of the EARLI study was to investigate the potential causes of autism by collecting environmental and biological data on 1000 mothers and their children with an Autism Spectrum Disorder (ASD) diagnosis to determine possible risk factors and biological indicators for autism during the prenatal, neonatal, and early postnatal periods. NIH researchers collected samples identical to MARBLES during the mother's pregnancy and, postnatally, from mothers, their child with an ASD, and the baby born during the study. The NIH team also gathered data from the child's medical records for 3 years after birth. EARLI was one of the only studies to include vaccination histories. The study had enrolled approximately 300 mothers after 2½ years when funding was suddenly discontinued. The three actual studies produced from EARLI samples looked at cord blood androgens, hormones in meconium and paternal sperm DNA methylation. We are not aware of any efforts to look at the vaccine data. Once again, all that effort and treasure has not produced usable results.

The Centers for Disease Control and Prevention (CDC) began the Study to Explore Early Development (SEED) in 2009.[7] The CDC touted SEED as one of the largest epidemiologic investigations of multiple genetic and environmental risk factors and causal pathways contributing to different ASD phenotypes. SEED promised to compare children aged 2–5 years with an autism spectrum disorder (ASD), to children from the general population and children with non-ASD developmental problems through parent-completed questionnaires, interviews, clinical evaluations, biospecimen sampling, and medical record abstraction with a focus on the prenatal and early postnatal periods. Researchers enrolled over 5,000 children in the study during two earlier phases. In 2016 the CDC announced an additional $27 million in funding to add a Phase 3 to SEED that will continue to enroll children until 2021. Altogether, SEED will enroll over 7,000 children. To date, eight years into the study, we know of 5 published papers based on SEED, none of which tests any hypothesis about autism causation.

Congress authorized the NIH to create the National Children's Study (NCS) under the Children's Health Act of 2000.[8] Congress charged the NIH to study environmental influences on child health and development. The NCS was to be a large-scale, long-term study of 100,000 US children from birth to age 21 and their parents. The NIH began the pilot study in 2009. Shortly after its inception, its Director, Dr. Duane Alexander, Director of the Eunice Kennedy Shriver National Institute of Child Health and Development (NICHD) publicly called for the inclusion of vaccines as a covariate in the study.[9] Shortly after this statement, Alexander transferred out of NICHD to an advisory position within the NIH.[10] By the time recruitment suddenly ended in July 2014,[11] the study had enrolled only 5000 children in 40 locations.[12] Of the 54 citations related to the NCS in PubMed, only seven studies actually attempted to investigate children's health. The other 47 describe the design, sample collection methods, recruitment methods, and challenges of doing the study. A debilitating lack of coordination and huge cost overruns dogged the NCS from its inception. The NCS samples are housed at NICHD.

NIH announced its launch of the Environmental Influences on Child Health Outcomes (ECHO) study in late 2016.[13] NIH press releases reported that the NIH would spend $157 million on the seven-year initiative.[14] ECHO will investigate how exposures to a range of environmental factors in early development, from conception through early childhood, influence the health of children and adolescents. NIH's press release quotes you as saying, "Every baby should have the best opportunity to remain healthy and thrive throughout childhood, ECHO will help us better understand the factors that contribute to optimal health in children." Your news release explained that "experiences during sensitive developmental windows, including around the time of conception, later in pregnancy, and during infancy and early childhood, can have long-lasting effects on the health of children. These experiences encompass a broad range of exposures, from air pollution and chemicals in our neighborhoods, to societal factors such as stress, to individual behaviors like sleep and diet. They may act through any number of biological processes, for example changes in the expression of genes or development of the immune system." Oddly, there is no mention of vaccines being a part of the investigation despite the fact that they are administered during these "sensitive developmental windows," have been found to alter neurodevelopment in animal models, and certainly affect expression of genes and the immune system. As you know, the Institute of Medicine (IOM) has repeatedly chided the NIH, the FDA and the CDC for failing to properly study the role of vaccines in this context.

ECHO promises to fund large existing pediatric cohorts with a goal of enrolling more than 50,000 children from diverse racial, geographic and socioeconomic backgrounds to conduct research on high-impact pediatric health outcomes. These cohort studies will analyze existing data as well as follow the children over time to address the early environmental origins of ECHO's health outcome areas. The ECHO health outcome areas are: Upper and lower airway, obesity, pre-, peri-, and postnatal outcomes, and neurodevelopment.[15]

While we appreciate ECHO's ambitious objectives, we are concerned about your glaring omission of vaccine data. We also worry that this is yet another long-term study that NIH will start and not follow through with. From the publicly available data, we estimate that the above-mentioned studies have or will cost the American taxpayer hundreds of millions of dollars. We do not believe that yet another study of this type is the most direct route to fast answers our country needs about the causes of America's devastating chronic disease epidemics.

As we stated during our meeting, we are requesting access to specific existing databases pertaining to vaccines and vaccine safety. These include the Vaccine Safety Datalink which houses the vaccine and health records of ten million children. As you know from the information provided by me and your own information searches, the CDC has made it extraordinarily difficult to conduct valid independent research in this database. Rather than duplicating previous efforts, we ask you use your clear authority as the statutory Chairman of the Interagency Task Force on Safer Childhood Vaccines to open up the VSD and to make available, to independent qualified scientists and the world's leading machine data analysis experts, the existing biological samples, questionnaires and medical records from previous longitudinal studies to investigate whether vaccines are associated with the epidemic of health disorders plaguing our children today. We are also requesting that you allow Google's medical records division, Verily, to automate the pathetic Vaccine Adverse Events Reporting System (VAERS) which now captures fewer than 1% of vaccine injuries.

Thank you for considering our requests.

Very truly yours,

Robert F. Kennedy Jr.
cc: Jared Kushner, Reed Cordish

APPENDIX D

Letter from Dr. Francis Collins, NIH Director, to Robert F. Kennedy Jr., 8/8/17[1]

 DEPARTMENT OF HEALTH & HUMAN SERVICES   Public Health Service

National Institutes of Health
Bethesda, Maryland 20892

August 8, 2017

Robert F. Kennedy, Jr.
Chairman of the Board
World Mercury Project
1227 North Peachtree Pkwy, Suite 202
Peachtree City, Georgia 30269

Dear Mr. Kennedy:

Thank you for your e-mail of June 21st and your letter of July 3rd, in follow up to our meeting at NIH on May 31, 2017. We have carefully reviewed and considered your questions and requests, but regretfully we are not optimistic that much further progress can be made in these discussions. The approaches that you and your colleagues have advocated rest on the assumption that vaccines are unsafe. In our view and that of the vast majority of objective experts in medicine and public health, there is overwhelming scientific evidence that supports the safety and exceptional value of vaccinations. While you have gone to some length to provide citations of studies that contradict that conclusion, most of those represent small and unreplicated studies published in less rigorous journals, and have clearly been chosen by you in a fashion that reflects your prior conclusion about vaccine risk (especially regarding thimerosal), not in a truly objective effort to discover the truth. Furthermore, you and your colleagues support your conclusions by equating legal decisions about vaccines as being equivalent to the results of carefully designed research studies. Very different criteria and standards are employed in these two different processes and a legal decision cannot substitute for a research finding.

Given these fundamental differences, we are at a loss for how to move these discussions forward productively. Compelling evidence documents that the benefit/risk ratio of vaccines is extremely high. We cannot condone studies that withhold vaccines from significant numbers of children, placing large segments of our population at risk for infectious diseases—some of which can be life-threatening—as has recently been demonstrated by the outbreaks of measles in Minnesota and Italy. The claimed causal connection between childhood vaccines and autism, based on an initial British study that has been conclusively shown to be fabricated, has been exhaustively studied in hundreds of thousands of children and resoundingly rejected.

We stand ready to re-engage if you are willing to reconsider your fundamental position, based on the weight of the evidence. We face many challenges to prevent disease. However, vaccines are not the problem, they are a solution.

Sincerely yours,

Francis S. Collins, M.D., Ph.D.
Director
National Institutes of Health

Lawrence A. Tabak, D.D.S., Ph.D.
Deputy Director
National Institutes of Health

Carrie D. Wolinetz, Ph.D.
Acting Chief of Staff and
Associate Director for Science Policy
National Institutes of Health

Diana W. Bianchi, M.D.
Director, *Eunice Kennedy Shriver* National
Institute of Child Health and Human
Development
National Institutes of Health

Linda S. Birnbaum, Ph.D.
Director, National Institute of
Environmental Health Sciences and
National Toxicology Program
National Institutes of Health

Anthony S. Fauci, M.D.
Director, National Institute of Allergy
and Infectious Diseases
National Institutes of Health

Joshua A. Gordon, M.D., Ph.D.
Director, National Institute of Mental Health
National Institutes of Health

cc: Jared Kushner
Senior Advisor to the President
The White House

Reed Cordish
Assistant to the President for Intragovernmental and Technology Initiatives
The White House

Mary-Sumpter Lapinski
Counselor to the Secretary for Public Health and Science
U.S. Department of Health and Human Services

US District Court Stipulation Shows HHS Is in Violation of the "Mandate for Safer Childhood Vaccines" as Specified in the National Childhood Vaccine Injury Act of 1986

HHS is in Violation of the "Mandate for Safer Childhood Vaccines" as Stipulated in the National Childhood Vaccine Injury Act of 1986[1,2]

UNITED STATES DISTRICT COURT
SOUTHERN DISTRICT OF NEW YORK

INFORMED CONSENT ACTION NETWORK,

                              Plaintiff,

        -against-

UNITED STATES DEPARTMENT OF HEALTH
AND HUMAN SERVICES

                              Defendant.

STIPULATION

18-cv-03215 (JMF)

WHEREAS, 42 U.S.C. § 300aa-27, entitled "Mandate for safer childhood vaccines,"

provides as follows:

(a) General rule

In the administration of this part and other pertinent laws under the
jurisdiction of the Secretary [of the Department of Health and Human
Services], the Secretary shall—

(1) promote the development of childhood vaccines that result in
fewer and less serious adverse reactions than those vaccines on the
market on December 22, 1987, and promote the refinement of such
vaccines, and

(2) make or assure improvements in, and otherwise use the
authorities of the Secretary with respect to, the licensing,
manufacturing, processing, testing, labeling, warning, use
instructions, distribution, storage, administration, field
surveillance, adverse reaction reporting, and recall of reactogenic
lots or batches, of vaccines, and research on vaccines, in order to
reduce the risks of adverse reactions to vaccines.

. . .

(c) Report

Within 2 years after December 22, 1987, and periodically thereafter,
the Secretary shall prepare and transmit to the Committee on Energy
and Commerce of the House of Representatives and the Committee on
Labor and Human Resources of the Senate a report describing the

1

Case 1:18-cv-03215-JMF   Document 18   Filed 07/09/18   Page 2 of 3

actions taken pursuant to subsection (a) of this section during the preceding 2-year period.

WHEREAS, on August 25, 2017, Informed Consent Action Network ("ICAN") submitted a Freedom of Information Act request (the "FOIA Request") to the Department of Health and Human Services ("HHS" or the "Department"), which was assigned control number 2017-01119-FOIA-OS, that sought the following records:

> **Any and all reports transmitted to the Committee on Energy and Commerce of the House of Representatives and the Committee on Labor and Human Resources of the Senate by the Secretary of HHS pursuant to 42 U.S.C. §300aa-27(c).**

WHEREAS, on April 12, 2018, ICAN filed a Complaint for Declaratory and Injunctive Relief in the United States District Court, Southern District of New York against HHS seeking records, if any, responsive to the FOIA Request;

WHEREAS, the HHS Immediate Office of the Secretary ("IOS") maintains the official correspondence file of the Secretary of HHS, including reports to Congress by the Secretary of HHS, and therefore those files were most likely to contain records responsive to the FOIA Request;

WHEREAS, on June 27, 2018, HHS sent ICAN the following response to the FOIA Request:

> The [Department]'s searches for records did not locate any records responsive to your request. The Department of Health and Human Services (HHS) Immediate Office of the Secretary (IOS) conducted a thorough search of its document tracking systems. The Department also conducted a comprehensive review of all relevant indexes of HHS Secretarial Correspondence records maintained at Federal Records Centers that remain in the custody of HHS. These searches did not locate records responsive to your request, or indications that records responsive to your request and in the custody of HHS are located at Federal Records Centers.

WHEREAS, ICAN believes the foregoing response from HHS now resolves all claims asserted in this action;

2

IT IS HEREBY STIPULATED AND AGREED, by and between the parties by and through their respective counsel:

1.   That the above-captioned action is voluntarily dismissed, with prejudice, pursuant to Federal Rule of Civil Procedure 41(a)(1)(A)(ii), each side to bear its own costs, attorney fees, and expenses; and

2.   That this stipulation may be signed in counterparts, and that electronic (PDF) signatures may be deemed originals for all purposes.

Dated: July 6, 2018   Dated: July 6, 2018
        New York, New York   New York, New York

        KENNEDY & MODONNA LLP   GEOFFREY S. BERMAN
        Attorney for Plaintiff   United States Attorney
                                                        Attorney for Defendant

By: _____   By: _____
        Robert F. Kennedy, Jr.   ANTHONY J. SUN
        48 Dewitt Mills Road   Assistant United States Attorney
        Hurley, NY 12443   86 Chambers Street, Third Floor
        (845) 481-2622   New York, New York 10007
                                                        (212) 637-2810
                                                        anthony.sun@usdoj.gov

                                                SO ORDERED:

                                                _____
                                                HON. JESSE M. FURMAN, U.S.D.J.

Dated: New York, New York   Any pending motions are moot. All conferences are
       July 6, 2018   vacated. The Clerk of Court is directed to close the case.

3

Endnotes

Foreword by Del Bigtree

1. "FEC Form 13, Reports of Accepted Donations For Inaugural Committee," Federal Election Commission, April 18, 2018, https://docquery.fec.gov/pdf/286/201704180300150286/201704180300150286.pdf, p. 163.
2. See Appendix E.

Chapter 1

1. "Vaccine History," The Children's Hospital of Philadelphia, accessed September 18, 2022, https://www.chop.edu/centers-programs/vaccine-education-center/vaccine-history/developments-by-year.
2. "Birth-18 Years Immunization Schedule," Centers for Disease Control and Prevention, accessed on September 15, 2022, https://www.cdc.gov/vaccines/schedules/hcp/imz/child-adolescent.html.
3. Nova, PBS, "Surviving AIDS," https://www.pbs.org/wgbh/nova/transcripts/2603aids.html, air date, February 2, 1999. Video link: https://www.youtube.com/watch?v=gpaUH5RK4eI&t=385s
4. US Food and Drug Administration, *ENGERIX-B: Package Insert*, US License No. 1617 (Research Triangle Park, NC: GlaxoSmithKline, 1989), https://www.fda.gov/media/119403/download.
5. US Food and Drug Administration, *INFANRIX: Package Insert*, US License No. 1617 (Research Triangle Park, NC: GlaxoSmithKline, 1997), https://www.fda.gov/media/75157/download.
6. US Food and Drug Administration, *ActHIB: Package Insert* (Swiftwater, PA: Sanofi Pasteur Inc., 1993), https://www.fda.gov/media/74395/download.
7. Kathleen Stratton et al., *Adverse Effects of Vaccines: Evidence and Causality* (Washington, DC: National Academies Press, 2011), doi:10.17226/13164.

[8]   Ibid.

[9]   Ibid.

[10] Ibid.

[11] Centers for Disease Control and Prevention, "Autism and Vaccines: Questions and Concerns," Vaccine Safety, accessed September 16, 2022, https://www .cdc.gov/vaccinesafety/concerns/autism.html.

[12] Committee on the Assessment of Studies of Health Outcomes Related to the Recommended Childhood Immunization Schedule, Board on Population Health and Public Health Practice, & Institute of Medicine, *The Childhood Immunization Schedule and Safety: Stakeholder Concerns, Scientific Evidence, and Future Studies* (Washington, DC: National Academies Press, 2013).

[13] Ibid. pg. 5.

[14] Ibid. pg. 6.

[15] Ibid. pg. 12.

[16] Ibid. pg. 14.

[17] A.J. Wakefield et al., "Ileal-Lymphoid-Nodular Hyperplasia, Non-Specific Colitis, and Pervasive Developmental Disorder in Children," *The Lancet* 351, no. 9103 (2018): 637–641. doi:10.1016/s0140-6736(97)11096-0.

[18] Tonya Bittner, "Wakefield'ed," *Urban Dictionary*, accessed on September 16, 2022, https://www.urbandictionary.com/define.php?term=Wakefield%27ed.

[19] Hannah Ritchie et al., "Coronavirus (COVID-19) Vaccinations," Our World in Data, accessed on April 15, 2023, https://ourworldindata.org/covid -vaccinations.

[20] MedAlerts.org, "Search the U.S. Government's VAERS Data," National Vaccine Information Center, accessed on April 15, 2023, https://medalerts.org /index.php.

[21] Anna Halkidis, "Vaccine Injuries Are Rare, Just Look at the Numbers," *Parents*, accessed September 12, 2022, https://www.parents.com/health /vaccines/vaccine-compensation-program-shows-vaccination-injuries-are -rare/.

[22] Fanny Wong, "Vaccine Injury Program Goes Unknown," ABA for Law Students, 2018, accessed September 12, 2022, https://abaforlawstudents.com /2016/04/11/the-largely-unknown-national-vaccine-injury-compensation -program/.

[23] Ross Lazarus et al., *Electronic Support for Public Health–Vaccine Adverse Event Reporting System (ESP: VAERS),* Grant ID: R18 HS 017045, Rockville, MD, The Agency for Healthcare Research and Quality (AHRQ), Mech2011, https://digital.ahrq.gov/sites/default/files/docs/publication/r18hs017045 -lazarus-final-report-2011.pdf.

[24] Adjuvants are substances used in combination with vaccine antigens to "produce a more robust immune response than the antigen alone."

Adjuvants stimulate cells in the innate immune system to "create a local immunocompetent environment at the injection site." Sunita Awate et al., "Mechanism of Action of Adjuvants," *Frontiers in Immunology* 4 (2013) 114, doi: 10.3389/fimmu.2013.00114.

25 FUTURE II Study Group, "Quadrivalent Vaccine Against Human Papillomavirus to Prevent High-Grade Cervical Lesions," *The New England Journal of Medicine* 356, no. 19 (2007): 1915–1927, doi:10.1056/ NEJMoa061741.

26 US Food and Drug Administration, *Gardasil 9: Package Insert*, USPI-v503-i-2008r012 (Whitehouse Station, NJ: Merck Sharp & Dohme Corp., 2020), https://www.fda.gov/media/90064/download.

27 Milagritos D. Tapia et al., "Maternal Immunisation with Trivalent Inactivated Influenza Vaccine for Prevention of Influenza in Infants in Mali: A Prospective, Active-Controlled, Observer-Blind, Randomised Phase 4 Trial," *The Lancet: Infectious Diseases* 16, no. 9 (2016): 1026-1035. doi:10.1016/ S1473-3099(16)30054-8.

28 The College of Physicians of Philadelphia, "Vaccines 101: Ethical Issues and Vaccines," The College of Physicians of Philadelphia, accessed September 19, 2022, https://cpp-hov.netlify.app/vaccines-101/ethical-issues-and-vaccines.

29 Food and Drug Administration, "Placebos and Blinding in Randomized Controlled Cancer Clinical Trials for Drug and Biological Products: Guidance for Industry," August 2019, https://www.fda.gov/media/130326/download.

30 Clovis Oncology, Inc., "A Study in Ovarian Cancer Patients Evaluating Rucaparib and Nivolumab as Maintenance Treatment Following Response to Front-Line Platinum-Based Chemotherapy (ATHENA)," (Clinicaltrials.gov Identifier NCT03522246), updated November 5, 2021, https://clinicaltrials .gov/ct2/show/NCT03522246.

31 American Regent Inc., 2021. "Randomized Placebo-controlled Trial of FCM as Treatment for Heart Failure with Iron Deficiency (HEART-FID)," (ClinicalTrials.gov Identifier: NCT03037931), updated November 16, 2021, https://clinicaltrials.gov/ct2/show/NCT03037931.

32 National Institute of Allergy and Infectious Diseases (NIAID), "Placebo-Controlled Trial of Antibiotic Therapy in Adults With Suspect Lower Respiratory Tract Infection (LRTI) and a Procalcitonin Level," (ClinicalTrials. gov Identifier: NCT03341273), updated August 24, 2021, https://www .clinicaltrials.gov/ct2/show/NCT03341273.

33 Priyanka Boghani, "Dr. Paul Offit: 'A Choice Not to Get a Vaccine Is Not a Risk-Free Choice,' *Frontline*," *Public Broadcasting Service*, November 20, 2015, https://www.pbs.org/wgbh/frontline/article/paul-offit-a-choice-not-to -get-a-vaccine-is-not-a-risk-free-choice/.

34 Ibid.

35 The College of Physicians of Philadelphia, "Vaccines 101: Ethical Issues and Vaccines," The College of Physicians of Philadelphia, accessed September 19, 2022, https://cpp-hov.netlify.app/vaccines-101/ethical-issues-and-vaccines.

36 The Cochrane Collaboration is an international network of researchers and health professionals headquartered in the UK and produces information for making healthcare decisions. They do not receive any commercial funding. Their mission is to be "an independent, diverse, global organization that collaborates to produce trusted synthesized evidence, make it accessible to all, and advocate for its use." https://www .cochrane.org/about-us accessed May 7, 2023.

37 Andrew Anglemyer et al., "Healthcare Outcomes Assessed with Observational Study Designs Compared with Those Assessed in Randomized Trials," *Cochrane Database of Systematic Reviews* (2014). doi:10.1002/14651858. MR000034.pub2.

38 Frank DeStefano et al., "Age at First Measles-Mumps-Rubella Vaccination in Children with Autism and School-Matched Control Subjects: A Population-Based Study in Metropolitan Atlanta," *Pediatrics* 113, no. 2 (2004): 259–266, doi:10.1542/peds.113.2.259.

39 Thomas Verstraeten et al., "Safety of Thimerosal-Containing Vaccines: A Two-Phased Study of Computerized Health Maintenance Organization Databases," *Pediatrics* 112, no. 5 (2003): 1039–1048. doi:10.1542/peds.112.5.1039.

40 Cristofer S. Price et al., "Prenatal and Infant Exposure to Thimerosal from Vaccines and Immunoglobulins and Risk of Autism," *Pediatrics* 126, no. 4 (2010): 656–664. doi:10.1542/peds.2010-0309.

41 Frank DeStefano et al., "Increasing Exposure to Antibody-Stimulating Proteins and Polysaccharides in Vaccines Is Not Associated with Risk of Autism," *The Journal of Pediatrics* 163, no. 2 (2013): 561–567. doi:10.1016/j. jpeds.2013.02.001.

42 Priyanka Boghani, "Dr. Paul Offit: 'A Choice Not to Get a Vaccine Is Not a Risk-Free Choice,' *Frontline*," *Public Broadcasting Service*, November 20, 2015, https://www.pbs.org/wgbh/frontline/article/paul-offit-a-choice-not-to -get-a-vaccine-is-not-a-risk-free-choice/.

43 Dan Olmsted, "The Age of Autism: The Amish Elephant," *UPI*, October 29, 2005, https://www.upi.com/Health_News/2005/10/29/The-Age-of-Autism -The-Amish-Elephant/44901130610898/.

44 Holly A. Hill et al., "Vaccination Coverage Among Children Aged 19–35 Months—United States, 2017," *Morbidity Mortality Weekly Report* 67, no. 40 (2018): 1123–1128. doi:10.15585/mmwr.mm6740a4.

45 "Amish in America, *American Experience*," *Public Broadcasting Service*, accessed July 15, 2022, https://www.pbs.org/wgbh/americanexperience /features/amish-in-america/.

46 Children's Health Defense, "Vaxxed-Unvaxxed: Parts I-XII," accessed on September 15, 2022, https://childrenshealthdefense.org/wp-content/uploads /Vaxxed-Unvaxxed-Parts-I-XII.pdf.

Chapter 2

1 Committee on the Assessment of Studies of Health Outcomes Related to the Recommended Childhood Immunization Schedule, Board on Population Health and Public Health Practice, & Institute of Medicine, *The Childhood Immunization Schedule and Safety; Stakeholder Concerns, Scientific Evidence, and Future Studies* (Washington, DC: National Academies Press, 2013).

2 Anthony R. Mawson, et al., "Pilot Comparative Study on the Health of Vaccinated and Unvaccinated 6- to 12-year-old U.S. Children," *Journal of Translational Science* 3, no. 3 (2017): 1-12, doi:10.15761/JTS.1000186.

3 Ibid.

4 Ibid.

5 Ibid.

6 Ibid.

7 Ibid.

8 Anthony R. Mawson et al., "Preterm Birth, Vaccination and Neurodevelopmental Disorders: A Cross-Sectional Study of 6- to 12-Year-Old Vaccinated and Unvaccinated Children," *Journal of Translational Science* 3, no. 3 (2017): 1-8, doi:10.15761/JTS.1000187.

9 Ibid.

10 Ibid

11 Committee on the Assessment of Studies of Health Outcomes Related to the Recommended Childhood Immunization Schedule, Board on Population Health and Public Health Practice, & Institute of Medicine, *The Childhood Immunization Schedule and Safety: Stakeholder Concerns, Scientific Evidence, and Future Studies* (Washington, DC: National Academies Press, 2013).

12 "Frontiers in Public Health," *Frontiers in Public Health*, accessed September 13, 2022, https://www.frontiersin.org/journals/public-health.

13 National Library of Medicine, *PubMed.gov.*, PubMed Overview, accessed September 13, 2022, https://pubmed.ncbi.nlm.nih.gov/about/.

14 COPE Council, *Guidelines: Retraction Guidelines—English*, Publicationetchics. org, November 19, 2019, https://www.medknow.com/documents/COPE%20 -Retraction%20Guidelines.pdf.

15 Ibid.

16 Anthony R. Mawson et al., "Preterm Birth, Vaccination and Neurodevelopmental Disorders: A Cross-Sectional Study of 6- to 12-Year-Old Vaccinated and Unvaccinated Children," *Journal of Translational Science* 3, no. 3 (2017): 1-8, doi:10.15761/JTS.1000187.

17 Anthony R. Mawson, et al., "Pilot Comparative Study on the Health of Vaccinated and Unvaccinated 6- to 12-year-old U.S. Children," *Journal of Translational Science* 3, no. 3 (2017): 1-12, doi:10.15761/JTS.1000186.

18 Brian Hooker and Neil Z. Miller, "Analysis of Health Outcomes in Vaccinated and Unvaccinated Children: Developmental Delays, Asthma, Ear Infections and Gastrointestinal Disorders," *SAGE Open Medicine* 8, (2020): 2050312120925344, doi:10.1177/2050312120925344.

19 Ibid.

20 Ibid.

21 Ibid.

22 Ibid.

23 Flora Teoh, "Significant Methodological Flaws in a 2020 Study Claiming to Show Unvaccinated Children are Healthier," Health Feedback, December 10, 2020, accessed September 13, 2022, https://healthfeedback.org/claimreview /significant-methodological-flaws-in-a-2020-study-claiming-to-show -unvaccinated-children-are-healthier-brian-hooker-childrens-health-defense/.

24 Brian Hooker and Neil Z. Miller, "Health Effects in Vaccinated versus Unvaccinated Children," *Journal of Translational Science* 7 (2021): 1–11, doi:10.15761/JTS.1000459.

25 Ibid.

26 Ibid.

27 Brian Hooker and Neil Z. Miller, "Analysis of Health Outcomes in Vaccinated and Unvaccinated Children: Developmental Delays, Asthma, Ear Infections and Gastrointestinal Disorders," *SAGE Open Medicine* 8, (2020): 2050312120925344, doi:10.1177/2050312120925344.

28 Brian Hooker and Neil Z. Miller, "Health Effects in Vaccinated versus Unvaccinated Children," *Journal of Translational Science* 7 (2021): 1–11, doi:10.15761/JTS.1000459.

29 Ibid.

30 Brian Hooker and Neil Z. Miller, "Analysis of Health Outcomes in Vaccinated and Unvaccinated Children: Developmental Delays, Asthma, Ear Infections and Gastrointestinal Disorders," *SAGE Open Medicine* 8, (2020): 2050312120925344, doi:10.1177/2050312120925344.

31 Brian Hooker and Neil Z. Miller, "Health Effects in Vaccinated versus Unvaccinated Children," *Journal of Translational Science* 7 (2021): 1–11, doi:10.15761/JTS.1000459.

32 Ibid.

33 Ibid.

34 Ibid.

35 Ibid.

36 James Lyons-Weiler and Paul Thomas, "Relative Incidence of Office Visits and Cumulative Rates of Billed Diagnoses along the Axis of Vaccination," *International Journal of Environmental Research and Public Health* 17, no. 22 (2020): 8674, doi:10.3390/ijerph17228674.

37 Ibid.

38 Ibid.

39 Ibid.

40 Ibid.

41 Ibid.

42 "Integrative Pediatrics: A Safe Passage in a Changing World," Integrative Pediatrics, accessed September 13, 2022, https://www.integrativepediatrics -online.com/.

43 Paul Thomas and Jennifer Margulis, *The Vaccine-Friendly Plan: Dr. Paul's Safe and Effective Approach to Immunity and Health-from Pregnancy through Your Child's Teen Years* (New York City: Ballantine Books, 2016).

44 Robert F. Kennedy Jr., "Join Me in Supporting Dr. Paul Thomas, a Hero Defending Children's Health," *The Defender*, December 17, 2020, https: //childrenshealthdefense.org/defender/support-dr-paul-thomas/.

45 Brian Hooker, interview with Dr. Paul Thomas, "Paul Thomas, and the Vaccine Friendly Plan," October 21, 2021, *Doctors and Scientists,* CHD.TV, https://live .childrenshealthdefense.org/shows/doctors-and-scientists-with-brian-hooker -phd/l8YY41rHQE.

46 Alix Mayer, "Groundbreaking Study Shows Unvaccinated Children Are Healthier than Vaccinated Children," *The Defender*, April 10, 2021, https: //childrenshealthdefense.org/defender/unvaccinated-children-healthier-than -vaccinated-children/.

47 Robert F. Kennedy Jr., "Join Me in Supporting Dr. Paul Thomas, a Hero Defending Children's Health," *The Defender*, December 17, 2020, https: //childrenshealthdefense.org/defender/support-dr-paul-thomas/.

48 Ibid.

49 Ibid.

50 "In the Matter of: Paul Norman Thomas, MD. License Number MD15689: Order of Emergency Suspension," Court Proceeding, Oregon Medical Board, 2020, https://omb.oregon.gov/Clients/ORMB/OrderDocuments/e579dd35 -7e1b-471f-a69a-3a800317ed4c.pdf.

51 Ibid.

52 Alix Mayer, "Groundbreaking Study Shows Unvaccinated Children Are Healthier than Vaccinated Children," *The Defender*, April 10, 2021, https: //childrenshealthdefense.org/defender/unvaccinated-children-healthier-than -vaccinated-children/.

53 Ibid.

54 Jeremy R. Hammond, *The War on Informed Consent: The Persecution of Dr. Paul Thomas by the Oregon Medical Board* (New York, NY: Skyhorse Publishing, 2021).

55 "In the Matter of: Paul Norman Thomas, MD. License Number MD15689: Interim Stipulated Order," Court Proceeding, Oregon Medical Board, 2021, https://omb.oregon.gov/Clients/ORMB/OrderDocuments/edf7724a-1cbb -46a7-a6c8-6f28fa2b337a.pdf.

56 Robert F. Kennedy Jr., "Join Me in Supporting Dr. Paul Thomas, a Hero Defending Children's Health," *The Defender*, December 17, 2020, https: //childrenshealthdefense.org/defender/support-dr-paul-thomas/.

57 "In the Matter of Paul Normal Thomas, MD, License Number MD15689: Stipulated Order," Court Proceeding, Oregon Medical Board, 2022, https: //omb.oregon.gov/clients/ormb/OrderDocuments/3f4010d3-92d5-43bb -bd1b-2c16b24260f0.pdf

58 International Journal of Environmental Research and Public Health Editorial Office, "Retraction: Lyons-Weiler, J.; Thomas, P. Relative Incidence of Office Visits and Cumulative Rates of Billed Diagnoses Along the Axis of Vaccination," *International Journal of Environmental Research and Public Health* 18, no. 15: 7754, doi:10.3390/ijerph18157754.

59 Ibid.

60 James Lyons-Weiler and Russell Blaylock, "Revisiting Excess Diagnoses of Illnesses and Conditions in Children Whose Parents Provided Informed Permission to Vaccinate Them," *International Journal of Vaccine Theory, Practice, and Research* 2, no. 2: 603-618, doi:10.56098/ijvtpr.v2i2.59.

61 NVKP, "Diseases and Vaccines: NVKP Survey Results," Nederlandse Vereniging Kritisch Prikken, 2006, accessed July 1, 2022, https://www.nvkp .nl/ziekten-en-vaccins/overzicht/enquete-2006/.

62 "Dutch National Immunisation Programme," accessed March 30, 2023, https://rijksvaccinatieprogramma.nl/english.

63 Ibid.

64 Ibid.

65 James Lyons-Weiler and Paul Thomas, "Relative Incidence of Office Visits and Cumulative Rates of Billed Diagnoses along the Axis of Vaccination," *International Journal of Environmental Research and Public Health* 17, no. 22 (2020): 8674, doi:10.3390/ijerph17228674.

66 Brian Hooker and Neil Z. Miller, "Health Effects in Vaccinated versus Unvaccinated Children," *Journal of Translational Science* 7, (2021): 1–11, doi:10.15761/JTS.1000459.

67 Brian Hooker and Neil Z. Miller, "Analysis of Health Outcomes in Vaccinated and Unvaccinated Children: Developmental Delays, Asthma, Ear

Infections and Gastrointestinal Disorders," *SAGE Open Medicine* 8, (2020): 2050312120925344, doi:10.1177/2050312120925344.

68 Anthony R. Mawson et al., "Pilot Comparative Study on the Health of Vaccinated and Unvaccinated 6- to 12-year-old U.S. Children," *Journal of Translational Science* 3, no. 3 (2017): 1-12, doi:10.15761/JTS.1000186.

69 "Dutch National Immunisation Programme," accessed March 30, 2023, https://rijksvaccinatieprogramma.nl/english.

70 Anthony R. Mawson et al., "Pilot Comparative Study on the Health of Vaccinated and Unvaccinated 6- to 12-year-old U.S. Children," *Journal of Translational Science* 3, no. 3 (2017): 1-12, doi:10.15761/JTS.1000186.

71 Brian Hooker and Neil Z. Miller, "Health Effects in Vaccinated versus Unvaccinated Children," *Journal of Translational Science* 7, (2021): 1–11, doi:10.15761/JTS.1000459.

72 James Lyons-Weiler and Paul Thomas, "Relative Incidence of Office Visits and Cumulative Rates of Billed Diagnoses along the Axis of Vaccination," *International Journal of Environmental Research and Public Health* 17, no. 22 (2020): 8674, doi:10.3390/ijerph17228674.

73 Ibid.

74 Brian Hooker and Neil Z. Miller, "Health Effects in Vaccinated versus Unvaccinated Children," *Journal of Translational Science* 7, (2021): 1–11, doi:10.15761/JTS.1000459.

75 Brian Hooker and Neil Z. Miller, "Analysis of Health Outcomes in Vaccinated and Unvaccinated Children: Developmental Delays, Asthma, Ear Infections and Gastrointestinal Disorders," *SAGE Open Medicine* 8, (2020): 2050312120925344, doi:10.1177/2050312120925344.

76 Brian Hooker and Neil Z. Miller, "Health Effects in Vaccinated versus Unvaccinated Children," *Journal of Translational Science* 7, (2021): 1–11, doi:10.15761/JTS.1000459.

77 Anthony R. Mawson et al., "Pilot Comparative Study on the Health of Vaccinated and Unvaccinated 6- to 12-year-old U.S. Children," *Journal of Translational Science* 3, no. 3 (2017): 1-12, doi:10.15761/JTS.1000186.

78 James Lyons-Weiler and Paul Thomas, "Relative Incidence of Office Visits and Cumulative Rates of Billed Diagnoses along the Axis of Vaccination," *International Journal of Environmental Research and Public Health* 17, no. 22 (2020): 8674, doi:10.3390/ijerph17228674.

79 Anthony R. Mawson et al., "Pilot Comparative Study on the Health of Vaccinated and Unvaccinated 6- to 12-year-old U.S. Children," *Journal of Translational Science* 3, no. 3 (2017): 1-12, doi:10.15761/JTS.1000186.

80 Joy Garner, "Statistical Evaluation of Health Outcomes in the Unvaccinated: Full Report," The Control Group: Pilot Survey of Unvaccinated Americans, November 19, 2020. https://truthpeep.com/wp-content/uploads/STATISTICAL

-EVALUATION-OF-HEALTH-OUTCOMES-IN-THE -UNVACCINATED.pdf.

[81] Ibid.

[82] Michael E. Rezaee and Martha Pollock, "Multiple Chronic Conditions among Outpatient Pediatric Patients, Southeastern Michigan, 2008–2013," *Preventing Chronic Disease* 12, (2015): E18, doi:10.5888/pcd12.140397.

[83] Carmela Avena-Woods, "Overview of Atopic Dermatitis," *American Journal of Managed Care* 23, no. 8 (2017): S115-S123. https://cdn.sanity .io/files/0vv8moc6/ajmc/e73485ac3035c1ff8cd31af0ba409136270ee250 .pdf/AJMC_ACE0068_06_2017_AtopicDermatitis_Overview_of_Atopic _Dermatitis.pdf.

[84] "Most Recent National Asthma Data," Asthma, Centers for Disease Control and Prevention, May 25, 2022, accessed September 16, 2022, https://www .cdc.gov/asthma/most_recent_national_asthma_data.htm.

[85] "Age-Adjusted Percentages (with Standard Errors) of Ever Having Been Told of Having a Learning Disability or Attention Deficit/ Hyperactivity Disorder for Children Aged 3–17 Years, by Selected Characteristics: United States, 2018," National Health Interview Survey, Centers for Disease Control and Prevention, 2018, accessed September 16, 2022, https://ftp.cdc.gov/pub/Health_Statistics/NCHS/NHIS/SHS/2018 _SHS_Table_C-3.pdf.

[86] "Data and Statistics about ADHD," Attention-Deficit/Hyperactivity Disorder (ADHD), Centers for Disease Control and Prevention, August 2, 2022, accessed September 16, 2022, https://www.cdc.gov/ncbddd/adhd /data.html.

[87] Benjamin Zablotsky et al., "Prevalence and Trends of Developmental Disabilities among Children in the United States: 2009–2017," *Pediatrics* 144, no.4 (2019): e20190811, doi:10.1542/peds.2019-0811.

[88] Lindsey I. Black et al., "Communication Disorders and Use of Intervention Services among Children Aged 3–17 Years: United States, 2012," NCHS Data Brief, No. 205 (Hyattsville, MD: National Center for Health Statistics, 2015), https://www.cdc.gov/nchs/products/databriefs/db205.htm.

[89] "Birth Defects," Centers for Disease Control and Prevention, August 29, 2022, accessed September 13, 2022, https://www.cdc.gov/ncbddd/birthdefects /index.html.

[90] Michael D. Kogan et al., "The Prevalence of Parent-Reported Autism Spectrum Disorder among US Children," *Pediatrics* 142, no. 6 (2018): e20174161, doi:10.1542/peds.2017-4161.

[91] Joy Garner, "Statistical Evaluation of Health Outcomes in the Unvaccinated: Full Report," The Control Group: Pilot Survey of Unvaccinated Americans, November 19, 2020. https://truthpeep.com/wp-content/uploads/STATISTICAL

-EVALUATION-OF-HEALTH-OUTCOMES-IN-THE -UNVACCINATED.pdf.

92 Ibid.

93 Brian Hooker and Neil Z. Miller, "Health Effects in Vaccinated versus Unvaccinated Children," *Journal of Translational Science* 7, (2021): 1–11, doi:10.15761/JTS.1000459.

94 Anthony R. Mawson et al., "Pilot Comparative Study on the Health of Vaccinated and Unvaccinated 6- to 12-year-old U.S. Children," *Journal of Translational Science* 3, no. 3 (2017): 1-12, doi:10.15761/JTS.1000186.

95 Rachel Enriquez et al., "The Relationship Between Vaccine Refusal and Self-Report of Atopic Disease in Children," *The Journal of Allergy and Clinical Immunology* 115, no. 4 (2005): 737-744, doi:10.1016/j.jaci.2004.12.1128.

96 Ibid.

97 Anthony R. Mawson et al., "Pilot Comparative Study on the Health of Vaccinated and Unvaccinated 6- to 12-year-old U.S. Children," *Journal of Translational Science* 3, no. 3 (2017): 1-12, doi:10.15761/JTS.1000186.

98 Brian Hooker and Neil Z. Miller, "Health Effects in Vaccinated versus Unvaccinated Children," *Journal of Translational Science* 7, (2021): 1–11, doi:10.15761/JTS.1000459.

99 NVKP, "Diseases and Vaccines: NVKP Survey Results," Nederlandse Vereniging Kritisch Prikken, 2006, accessed July 1, 2022, https://www.nvkp .nl/ziekten-en-vaccins/overzicht/enquete-2006/.

100 Rachel Enriquez et al., "The Relationship Between Vaccine Refusal and Self-Report of Atopic Disease in Children," *The Journal of Allergy and Clinical Immunology* 115, no. 4 (2005): 737–744, doi:10.1016/j.jaci.2004.12.1128.

101 Matthew F. Dalcy et al., "Association Between Aluminum Exposure from Vaccines Before Age 24 Months and Persistent Asthma at Age 24 to 59 Months," *Academic Pediatrics* 23, no. 1 (2023); 37–46, doi:10.1016/j. acap.2022.08.006.

102 Ibid.

103 Ibid.

104 Ibid.

105 Rachel Enriquez et al., "The Relationship Between Vaccine Refusal and Self-Report of Atopic Disease in Children," *The Journal of Allergy and Clinical Immunology* 115, no. 4 (2005): 737–744, doi:10.1016/j.jaci.2004.12.1128.

106 Joy Garner, "Statistical Evaluation of Health Outcomes in the Unvaccinated: Full Report," The Control Group: Pilot Survey of Unvaccinated Americans, November 19, 2020. https://truthpeep.com/wp-content/uploads/STATISTICAL -EVALUATION-OF-HEALTH-OUTCOMES-IN-THE- UNVACCINATED.pdf.

[107] NVKP, "Diseases and Vaccines: NVKP Survey Results," Nederlandse Vereniging Kritisch Prikken, 2006, accessed July 1, 2022, https://www.nvkp .nl/ziekten-en-vaccins/overzicht/enquete-2006/.

[108] James Lyons-Weiler and Paul Thomas, "Relative Incidence of Office Visits and Cumulative Rates of Billed Diagnoses along the Axis of Vaccination," *International Journal of Environmental Research and Public Health* 17, no. 22 (2020): 8674, doi:10.3390/ijerph17228674.

[109] Brian Hooker and Neil Z. Miller, "Health Effects in Vaccinated versus Unvaccinated Children," *Journal of Translational Science* 7, (2021): 1–11, doi:10.15761/JTS.1000459.

[110] Brian Hooker and Neil Z. Miller, "Analysis of Health Outcomes in Vaccinated and Unvaccinated Children: Developmental Delays, Asthma, Ear Infections and Gastrointestinal Disorders," *SAGE Open Medicine* 8, (2020): 2050312120925344, doi:10.1177/2050312120925344.

[111] James Lyons-Weiler and Paul Thomas, "Relative Incidence of Office Visits and Cumulative Rates of Billed Diagnoses along the Axis of Vaccination," *International Journal of Environmental Research and Public Health* 17, no. 22 (2020): 8674, doi:10.3390/ijerph17228674.

[112] NVKP, "Diseases and Vaccines: NVKP Survey Results," Nederlandse Vereniging Kritisch Prikken, 2006, accessed July 1, 2022, https://www.nvkp .nl/ziekten-en-vaccins/overzicht/enquete-2006/.

[113] Annika Klopp et al., "Modes of Infant Feeding and the Risk of Childhood Asthma: A Prospective Birth Cohort Study," *The Journal of Pediatrics* 190, (2017): 192-199.e2. doi:10.1016/j.jpeds.2017.07.012.

[114] NVKP, "Diseases and Vaccines: NVKP Survey Results," Nederlandse Vereniging Kritisch Prikken, 2006, accessed July 1, 2022, https://www.nvkp .nl/ziekten-en-vaccins/overzicht/enquete-2006/.

Chapter 3

[1] David Kirby, *Evidence of Harm Mercury in Vaccines and the Autism Epidemic: A Medical Controversy* (New York, NY: St. Martin's Griffin, 2005).

[2] Robert F. Kennedy Jr. et al., *Thimerosal: Let the Science Speak: The Evidence Supporting the Immediate Removal of Mercury—a Known Neurotoxin—from Vaccines* (New York, NY: Skyhorse, 2015).

[3] *Trace Amounts: Autism, Mercury, and the Hidden Truth*, directed by Eric Gladen and Shiloh Levine (West Hollywood, CA: Gathr Films, 2014), DVD.

[4] "Historical Development of the Mercury Based Preservative Thimerosal," Children's Health Defense, accessed September 12, 2022, https://childrenshealthdefense.org/known-culprits/mercury/thimerosal-history/.

5   David A. Geier et al., "Thimerosal: Clinical, Epidemiologic and Biochemical
    Studies," *Clinica Chimica Acta*, 444 (2015): 212–20, doi:10.1016/j.
    cca.2015.02.030.

6   Neal A. Halsey, "Limiting Infant Exposure to Thimerosal in Vaccines and
    Other Sources of Mercury," *Journal of the American Medical Association* 282,
    no. 18 (1999): 1763–1766, doi:10.1001/jama.282.18.1763.

7   Put Children First, "Thimerosal Timeline," accessed on September 19, 2022,
    https://childrenshealthdefense.org/wp-content/uploads/THIMEROSAL
    -TIMELINE-PRE-1999-TO-2004.pdf.

8   Ibid.

9   Thomas M. Verstraeten et al., "Increased Risk of Developmental Neurological
    Impairment After High Exposure to Thimerosal-Containing Vaccine in First
    Month of Life," Epidemic Intelligence Service, accessed on September 24, 2022,
    https://childrenshealthdefense.org/wp-content/uploads/1999-eis-conference
    -abstract-presentation-verstraeten-et-al.pdf.

10 Ibid.

11 Ibid.

12 Ibid.

13 Ibid.

14 Thomas M. Verstraeten et al., "Scientific Review of Vaccine Safety Datalink
    Information," (transcript, Simpsonwood Retreat Center, Norcross, Georgia,
    June 7–8, 2000), https://www.putchildrenfirst.org/media/2.9.pdf.

15 Ibid.

16 Ibid.

17 Thomas M. Verstraeten et al., "Safety of Thimerosal-Containing Vaccines:
    A Two-Phased Study of Computerized Health Maintenance Organization
    Databases," *Pediatrics* 112, no. 5 (2003): 1039–1048, doi:10.1542/peds.
    112.5.1039.

18 Ibid.

19 Thomas M. Verstraeten "Thimerosal, the Centers for Disease Control and
    Prevention, and GlaxoSmithKline," *Pediatrics* 113, no. 4 (2004): 932,
    doi:10.1542./peds.113.4.932.

20 Institute of Medicine (US) Immunization Safety Review Committee,
    Immunization Safety Review: Vaccines and Autism (Washington, DC: National
    Academies Press, 2004), doi:10.17226/10997.

21 Brian Hooker et al., "Methodological Issues and Evidence of Malfeasance
    in Research Purporting to Show Thimerosal in Vaccines is Safe," *BioMed
    Research International* 2014, (2014): 247218, doi:10.1155/2014/247218.

22 Kreesten M. Madsen et al. 2003. "Thimerosal and the Occurrence of Autism:
    Negative Ecological Evidence from Danish Population-Based Data," *American
    Academy of Pediatrics* 112, no. 3 Pt 1: 604-606. doi:10.1542/peds.112.3.604.

23 Ibid.

24 Put Children First, "Thimerosal Timeline," accessed on September 19, 2022, https://childrenshealthdefense.org/wp-content/uploads/THIMEROSAL -TIMELINE-PRE-1999-TO-2004.pdf.

25 Ibid.

26 Ibid.

27 Heather A. Young, David A. Geier, and Mark R. Geier, "Thimerosal Exposure in Infants and Neurodevelopmental Disorders: An Assessment of Computerized Medical Records in the Vaccine Safety Datalink," *Journal of the Neurological Sciences* 271, no. 1–2 (2008): 110–118. doi:10.1016/j. jns.2008.04.002.

28 Ibid.

29 David A. Geier et al., "A Two-Phase Study Evaluating the Relationship Between Thimerosal-Containing Vaccine Administration and the Risk for an Autism Spectrum Disorder Diagnosis in the United States," *Translational Neurodegeneration* 2, no. 1 (2013): 25, doi:10.1186/2047-9158-2-25.

30 David A. Geier et al., "Thimerosal-Containing Hepatitis B Vaccination and the Risk for Diagnosed Specific Delays in Development in the United States: A Case-Control Study in the Vaccine Safety Datalink," *North American Journal of Medical Sciences* 6, no. 10 (2014): 519–531, doi:10.4103/1947-2714.143284.

31 David A. Geier et al., "Thimerosal Exposure and Disturbance of Emotions Specific to Childhood and Adolescence: A Case-Control Study in the Vaccine Safety Datalink (VSD) Database," *Brain Injury* 31, no. 2 (2017): 272–278. doi:10.1080/02699052.2016.1250950.

32 David A. Geier, Janet K. Kern, and Mark R. Geier, "Premature Puberty and Thimerosal-Containing Hepatitis B Vaccination: A Case-Control Study in the Vaccine Safety Datalink," *Toxics* 6, no. 4 (2018): 67, doi:10.3390/ toxics6040067.

33 David A. Geier et al., "Thimerosal: Clinical, Epidemiologic and Biochemical Studies," *Clinica Chimica Acta*, 444 (2015): 212–20, doi:10.1016/j.cca. 2015.02.030.

34 Put Children First, "Thimerosal Timeline," accessed on September 19, 2022, https://childrenshealthdefense.org/wp-content/uploads/THIMEROSAL -TIMELINE-PRE-1999-TO-2004.pdf.

35 Heather A. Young, David A. Geier, and Mark R. Geier, "Thimerosal Exposure in Infants and Neurodevelopmental Disorders: An Assessment of Computerized Medical Records in the Vaccine Safety Datalink," *Journal of the Neurological Sciences* 271, no. 1–2 (2008): 110–118. doi:10.1016/j. jns.2008.04.002.

36 Ibid.

[37] David A. Geier et al., "A Two-Phase Study Evaluating the Relationship Between Thimerosal-Containing Vaccine Administration and the Risk for an Autism Spectrum Disorder Diagnosis in the United States," *Translational Neurodegeneration* 2, no. 1 (2013): 25, doi:10.1186/2047-9158-2-25.

[38] Ibid.

[39] Ibid.

[40] Ibid.

[41] Ibid.

[42] David A. Geier et al., "Thimerosal-Containing Hepatitis B Vaccination and the Risk for Diagnosed Specific Delays in Development in the United States: A Case-Control Study in the Vaccine Safety Datalink," *North American Journal of Medical Sciences* 6, no. 10 (2014): 519–531, doi:10.4103/1947-2714.143284.

[43] David A. Geier et al., "Thimerosal Exposure and Increased Risk for Diagnosed Tic Disorder in the United States: A Case-Control Study," *Interdisciplinary Toxicology* 8, no. 2 (2015): 68–76, doi: 10.1515/intox-2015-0011.

[44] David A. Geier et al., "Thimerosal-Containing Hepatitis B Vaccination and the Risk for Diagnosed Specific Delays in Development in the United States: A Case-Control Study in the Vaccine Safety Datalink," *North American Journal of Medical Sciences* 6, no. 10 (2014): 519–531, doi:10.4103/1947-2714.143284.

[45] David A. Geier et al., "Thimerosal Exposure and Increased Risk for Diagnosed Tic Disorder in the United States: A Case-Control Study," Interdisciplinary Toxicology 8, no. 2 (2015): 68–76, doi:10.1515/intox-2015-0011.

[46] David A. Geier et al., "Thimerosal-Containing Hepatitis B Vaccination and the Risk for Diagnosed Specific Delays in Development in the United States: A Case-Control Study in the Vaccine Safety Datalink," *North American Journal of Medical Sciences* 6, no. 10 (2014): 519–531, doi:10.4103/1947-2714.143284.

[47] David A. Geier et al., "Thimerosal Exposure and Increased Risk for Diagnosed Tic Disorder in the United States: A Case-Control Study," Interdisciplinary Toxicology 8, no. 2 (2015): 68–76, doi:10.1515/intox-2015-0011.

[48] Ibid.

[49] David A. Geier et al., "Thimerosal-Containing Hepatitis B Vaccination and the Risk for Diagnosed Specific Delays in Development in the United States: A Case-Control Study in the Vaccine Safety Datalink," North American Journal of Medical Sciences 6, no. 10 (2014): 519–531, doi:10.4103/1947-2714.143284.

[50] David A. Geier et al., "Thimerosal Exposure and Disturbance of Emotions Specific to Childhood and Adolescence: A Case-Control Study in the Vaccine

Safety Datalink (VSD) Database," *Brain Injury* 31, no. 2 (2017): 272–278. doi:10.1080/02699052.2016.1250950.

51 David A. Geier, Janet K. Kern, and Mark R. Geier, "Premature Puberty and Thimerosal-Containing Hepatitis B Vaccination: A Case-Control Study in the Vaccine Safety Datalink," *Toxics* 6, no. 4 (2018): 67, doi:10.3390/toxics6040067.

52 David A. Geier et al., "Thimerosal Exposure and Disturbance of Emotions Specific to Childhood and Adolescence: A Case-Control Study in the Vaccine Safety Datalink (VSD) Database," *Brain Injury* 31, no. 2 (2017): 272–278. doi:10.1080/02699052.2016.1250950.

53 David A. Geier, Janet K. Kern, and Mark R. Geier, "Premature Puberty and Thimerosal-Containing Hepatitis B Vaccination: A Case-Control Study in the Vaccine Safety Datalink," *Toxics* 6, no. 4 (2018): 67, doi:10.3390/toxics6040067.

54 Ibid.

55 Ibid.

56 David A. Geier et al., "Thimerosal Exposure and Disturbance of Emotions Specific to Childhood and Adolescence: A Case-Control Study in the Vaccine Safety Datalink (VSD) Database," *Brain Injury* 31, no. 2 (2017): 272–278. doi:10.1080/02699052.2016.1250950.

57 Paul E. M. Fine and Robert T. Chen, "Confounding in Studies of Adverse Reactions to Vaccines," *American Journal of Epidemiology* 136, no. 2 (1992): 121–135, doi:10.1093/oxfordjournals.aje.a116479.

58 David A. Geier et al., "Thimerosal-Containing Hepatitis B Vaccination and the Risk for Diagnosed Specific Delays in Development in the United States: A Case-Control Study in the Vaccine Safety Datalink," *North American Journal of Medical Sciences* 6, no. 10 (2014): 519–531, doi:10.4103/1947-2714.143284.

59 Carolyn M. Gallagher and Melody S. Goodman, "Hepatitis B Vaccination of Male Neonates and Autism Diagnosis, NHIS 1997–2002," *Journal of Toxicology and Environmental Health Part A* 73, no. 24 (2010): 1665–1677, doi:10.1080/15287394.2010.519317.

60 Ibid.

61 Ibid.

62 Ibid.

63 Carolyn M. Gallagher and Melody S. Goodman, "Hepatitis B Triple Series Vaccine and Developmental Disability in US Children Aged 1–9 Years," *Toxicology and Environmental Chemistry* 90, no. 5 (2008): 997–1008, doi:10.1080/02772240701806501.

64 Ibid.

65 Ibid.

66 Thomas M. Verstraeten et al., "Increased Risk of Developmental Neurological Impairment After High Exposure to Thimerosal-Containing Vaccine in First Month of Life," Epidemic Intelligence Service, accessed on September 24, 2022, https://childrenshealthdefense.org/wp-content/uploads/1999-eis-conference -abstract-presentation-verstraeten-et-al.pdf.

67 William W. Thompson et al., "Early Thimerosal Exposure and Neuropsychological Outcomes at 7 to 10 Years," *The New England Journal of Medicine* 357, no. 13 (2007): 1281–1292. doi:10.1056/NEJMoa071434.

68 Ibid.

69 Ibid.

70 Ibid.

71 Ibid.

72 Ibid.

73 Ibid.

74 John P. Barile et al., "Thimerosal Exposure in Early Life and Neuropsychological Outcomes 7–10 Years Later," *Journal of Pediatric Psychology* 37, no. 1 (2012): 106–118. doi:10.1093/jpepsy/jsr048.

75 Nick Andrews et al., "Thimerosal Exposure in Infants and Developmental Disorders: A Retrospective Cohort Study in the United Kingdom Does Not Support a Causal Association," *Pediatrics* 114, no. 3 (2004): 584–591, doi:10.1542/peds.2003-1177-L.

76 Ibid.

77 William W. Thompson et al., "Early Thimerosal Exposure and Neuropsychological Outcomes at 7 to 10 Years," *The New England Journal of Medicine* 357, no. 13 (2007): 1281–1292. doi:10.1056/NEJMoa071434.

78 Ibid.

79 Ibid.

80 Nick Andrews et al., "Thimerosal Exposure in Infants and Developmental Disorders: A Retrospective Cohort Study in the United Kingdom Does Not Support a Causal Association," *Pediatrics* 114, no. 3 (2004): 584–591, doi:10.1542/peds.2003-1177-L.

81 Ibid.

82 Ibid.

83 Thomas M. Verstraeten et al., "Increased Risk of Developmental Neurological Impairment After High Exposure to Thimerosal-Containing Vaccine in First Month of Life," Epidemic Intelligence Service, accessed on September 24, 2022, https://childrenshealthdefense.org/wp-content/uploads/1999-eis -conference-abstract-presentation-verstraeten-et-al.pdf.

84 Ibid.

85 Ibid.

86 US Food and Drug Administration, *FLUVIRIN®: Package Insert*, 2007–2018 formulation (Summit, NJ: Seqirus, Inc., updated 2017), https://www.fda .gov/files/vaccines%2C%20blood%20%26%20biologics/published /Package-Insert—Fluvirin.pdf.

87 Centers for Disease Control and Prevention, "Seasonal Influenza Vaccine Supply for the U.S. 2022-2023 Influenza Season," accessed on April 13, 2023, https://www.cdc.gov/flu/prevent/vaxsupply.htm.

88 Pan American Health Organization, "Health in the Minamata Convention on Mercury," accessed on April 13, 2023, https://www3.paho.org/hq/index .php?option=com_content&view=article&id=8162:2013-health-minamata -convention-on-mercury&Itemid=0&lang=en#gsc.tab=0.

89 Carolyn M. Gallagher and Melody S. Goodman. 2010. "Hepatitis B Vaccination of Male Neonates and Autism Diagnosis, NHIS 1997-2002," *Journal of Toxicology and Environmental Health Part A* 73, no. 24 (2010): 1665-167, doi:10.1080/15287394.2010.519317.

90 David A. Geier, Janet K. Kern and Mark R. Geier, "Premature Puberty and Thimerosal-Containing Hepatitis B Vaccination: A Case-Control Study in the Vaccine Safety Datalink," *Toxics* 6, no. 4 (2018): 67, doi:10.3390/ toxics6040067.

91 David A. Geier et al., "Thimerosal Exposure and Disturbance of Emotions Specific to Childhood and Adolescence: A Case-Control Study in the Vaccine Safety Datalink (VSD) Database," *Brain Injury* 31, no. 2 (2017): 272-278. do i:10.1080/02699052.2016.1250950.

92 David A. Geier et al., "Thimerosal-Containing Hepatitis B Vaccination and the Risk for Diagnosed Specific Delays in Development in the United States: A Case-Control Study in the Vaccine Safety Datalink," *North American Journal of Medical Sciences* 6, no. 10 (2014): 519-531, doi:10.4103/1947-2714.143284.

93 David A. Geier et al., "A Two-Phase Study Evaluating the Relationship Between Thimerosal-Containing Vaccine Administration and the Risk for an Autism Spectrum Disorder Diagnosis in the United States," *Translational Neurodegeneration* 2, no. 1 (2013): 25, doi:10.1186/2047-9158-2-25.

94 Heather A. Young, David A. Geier and Mark R. Geier, "Thimerosal Exposure in Infants and Neurodevelopmental Disorders: An Assessment of Computerized Medical Records in the Vaccine Safety Datalink," *Journal of the Neurological Sciences* 271, no. 1-2 (2008): 110-118. doi:10.1016/j.jns.2008.04.002.

95 Thomas M. Verstraeten et al., "Increased Risk of Developmental Neurological Impairment After High Exposure to Thimerosal-Containing Vaccine in First Month of Life," Epidemic Intelligence Service, accessed on September 24, 2022, https://childrenshealthdefense.org/wp-content/uploads/1999-eis -conference-abstract-presentation-verstraeten-et-al.pdf.

96 Heather A. Young, David A. Geier and Mark R. Geier, "Thimerosal Exposure in Infants and Neurodevelopmental Disorders: An Assessment of Computerized Medical Records in the Vaccine Safety Datalink," *Journal of the Neurological Sciences* 271, no. 1-2 (2008): 110-118. doi:10.1016/j.jns.2008.04.002.

97 Ibid.

98 David A. Geier et al., "Thimerosal Exposure and Increased Risk for Diagnosed Tic Disorder in the United States: A Case-Control Study," Interdisciplinary Toxicology 8, no. 2 (2015): 68–76, doi:10.1515/intox-2015-0011.

99 William W. Thompson et al., "Early Thimerosal Exposure and Neuropsychological Outcomes at 7 to 10 Years," *The New England Journal of Medicine* 357, no. 13 (2007): 1281–1292. doi:10.1056/NFJMoa071434.

100 Nick Andrews et al., "Thimerosal Exposure in Infants and Developmental Disorders: A Retrospective Cohort Study in the United Kingdom Does Not Support a Causal Association," *Pediatrics* 114, no. 3 (2004): 584–591, doi:10.1542/peds.2003-1177-L.

101 William W. Thompson et al., "Early Thimerosal Exposure and Neuropsychological Outcomes at 7 to 10 Years," *The New England Journal of Medicine* 357, no. 13 (2007): 1281–1292. doi:10.1056/NEJMoa071434.

102 Carolyn M. Gallagher and Melody S. Goodman, 2010, "Hepatitis B Vaccination of Male Neonates and Autism Diagnosis, NHIS 1997–2002," *Journal of Toxicology and Environmental Health Part A* 73, no. 24 (2010): 1665–1677, doi:10.1080/15287394.2010.519317.

103 David A. Geier, Janet K. Kern and Mark R. Geier, "Premature Puberty and Thimerosal-Containing Hepatitis B Vaccination: A Case-Control Study in the Vaccine Safety Datalink," *Toxics* 6, no. 4 (2018): 67, doi:10.3390/toxics6040067.

104 David A. Geier et al., "Thimerosal Exposure and Disturbance of Emotions Specific to Childhood and Adolescence: A Case-Control Study in the Vaccine Safety Datalink (VSD) Database," *Brain Injury* 31, no. 2 (2017): 272–278. doi:10.1080/02699052.2016.1250950.

105 David A. Geier et al., "Thimerosal-Containing Hepatitis B Vaccination and the Risk for Diagnosed Specific Delays in Development in the United States: A Case-Control Study in the Vaccine Safety Datalink," *North American Journal of Medical Sciences* 6, no. 10 (2014): 519–531, doi:10.4103/1947-2714.143284.

106 David A. Geier et al., "A Two-Phase Study Evaluating the Relationship Between Thimerosal-Containing Vaccine Administration and the Risk for an Autism Spectrum Disorder Diagnosis in the United States," *Translational Neurodegeneration* 2, no. 1 (2013): 25, doi:10.1186/2047-9158-2-25.

107 David A. Geier, Janet K. Kern, and Mark R. Geier, "Premature Puberty and Thimerosal-Containing Hepatitis B Vaccination: A Case-Control Study

in the Vaccine Safety Datalink," *Toxics* 6, no. 4 (2018): 67, doi:10.3390/toxics6040067.

[108] David A. Geier et al., "Thimerosal Exposure and Disturbance of Emotions Specific to Childhood and Adolescence: A Case-Control Study in the Vaccine Safety Datalink (VSD) Database," *Brain Injury* 31, no. 2 (2017): 272–278. doi:10.1080/02699052.2016.1250950.

[109] David A. Geier et al., "Thimerosal-Containing Hepatitis B Vaccination and the Risk for Diagnosed Specific Delays in Development in the United States: A Case-Control Study in the Vaccine Safety Datalink," *North American Journal of Medical Sciences* 6, no. 10 (2014): 519–531, doi:10.4103/1947-2714.143284.

[110] David A. Geier et al., "A Two-Phase Study Evaluating the Relationship Between Thimerosal-Containing Vaccine Administration and the Risk for an Autism Spectrum Disorder Diagnosis in the United States," *Translational Neurodegeneration* 2, no. 1 (2013): 25, doi:10.1186/2047-9158-2-25.

Chapter 4

[1] Andrew J. Wakefield et al., "Ileal-Lymphoid-Nodular Hyperplasia, Non-Specific Colitis, and Pervasive Developmental Disorder in Children," *The Lancet* 351, no. 9103 (1998): 637–641, doi:10.1016/S0140-6736(97)11096-0.

[2] Andrew J. Wakefield, *Callous Regard: Autism and Vaccines—The Truth Behind a Tragedy* (New York: Skyhorse Publishing, 2017), ISBN: 9781510729667.

[3] Frank DeStefano et al., "Age at First Measles-Mumps-Rubella Vaccination in Children with Autism and School-Matched Control Subjects: A Population-Based Study in Metropolitan Atlanta," *Pediatrics* 113, no. 2 (2004): 259–266, doi:10.1542/peds.113.2.259.

[4] Ibid.

[5] Ibid.

[6] Ibid.

[7] Brian S. Hooker, "Reanalysis of CDC Data on Autism Incidence and Time of First MMR Vaccination," *Journal of American Physicians and Surgeons* 23, no. 4 (2018): 105–109, https://www.jpands.org/vol23no4/hooker.pdf.

[8] Ibid.

[9] Ibid.

[10] Ibid.

[11] Frank DeStefano et al., "Age at First Measles-Mumps-Rubella Vaccination in Children with Autism and School-Matched Control Subjects: A Population-Based Study in Metropolitan Atlanta," *Pediatrics* 113, no. 2 (2004): 259–266, doi:10.1542/peds.113.2.259.

12 Nick P. Thompson et al., "Is Measles Vaccination a Risk Factor for Inflammatory Bowel Disease?," *The Lancet* 345, no. 8947 (1995): 1071–1074, doi:10.1016/S0140-6736(95)90816-1.

13 Ibid.

14 Seif O. Shaheen et al., "Measles and Atopy in Guinea-Bissau," *The Lancet* 347, no. 9018 (1996): 1792–1796, doi:10.1016/s0140-6736(96)91617-7.

15 Ibid.

16 American Academy of Allergy, Asthma and Immunology, "Atopy Defined," accessed May 17, 2023, https://www.aaaai.org/tools-for-the-public/allergy-asthma-immunology-glossary/atopy-defined.

17 John Barthelow Classen, "Risk of Vaccine Induced Diabetes in Children with a Family History of Type 1 Diabetes," *The Open Pediatric Medicine Journal* 2, (2008): 7–10, doi:10.2174/1874309900802010007.

18 Ibid.

19 Ibid.

20 Guillaume Pineton de Chambrun et al., "Vaccination and Risk for Developing Inflammatory Bowel Disease: A Meta-Analysis of Case-Control and Cohort Studies," *Clinical Gastroenterology and Hepatology* 13, no. 8 (2015): 1405–1415 e1, doi:10.1016/j.cgh.2015.04.179.

21 Ibid.

22 Manish M. Patel et al., "Intussusception Risk and Health Benefits of Rotavirus Vaccination in Mexico and Brazil," *The New England Journal of Medicine* 364, (2011): 2283–2292, doi:10.1056/NEJMoa1012952.

23 Ibid.

24 Children's Hospital of Philadelphia, "Intussusception," accessed on April 13, 2023, https://www.chop.edu/conditions-diseases/intussusception.

25 Medscape, "Intussusception," accessed on April 13, 2023, https://emedicine.medscape.com/article/930708-overview.

26 US Food and Drug Administration, Rotarix®: Package Insert, US License 1617 (Triangle Park, NC: GlaxoSmithKline, 2008), https://www.fda.gov/media/75726/download.

27 Priya Kassin and Guy D. Eslick, "Risk of Intussusception Following Rotavirus Vaccination: An Evidence Based Meta-Analysis of Cohort and Case-Control Studies," *Vaccine* 35, no. 33 (2017): 4276–4286, doi:10.1016/j.vaccine.2017.05.064.

28 Ibid.

29 US Food and Drug Administration, RotaTeq®: Package Insert, STN: BL125122 (Whitehouse Station, NJ: Merck & Co. Inc., 2006), https://www.fda.gov/media/75718/download.

30 Children's Hospital of Philadelphia, "Intussusception," accessed on April 13, 2023, https://www.chop.edu/conditions-diseases/intussusception.

31 Medscape, "Intussusception," accessed on April 13, 2023, https://emedicine
    .medscape.com/article/930708-overview.
32 US Centers for Disease Control and Prevention, "Rotavirus Vaccine
    (Rotashield®) and Intussusception," accessed on April 13, 2023, https://www
    .cdc.gov/vaccines/vpd-vac/rotavirus/vac-rotashield-historical.htm.

Chapter 5

1   Lynette Luria and Gabriella Cardoza-Favarato, *Human Papillomavirus*
    (Treasure Island: Stat Pearls Publishing, 2022), Bookshelf ID: NBK448132.
2   Ibid.
3   The American College of Obstetricians and Gynecologists, "Loop
    Electrosurgical Excision Procedure (LEEP)," updated February, 2022, https:
    //www.acog.org/womens-health/faqs/loop-electrosurgical-excision-procedure.
4   US Food and Drug Administration, "June 8, 2006 Approval Letter—
    Human Papillomavirus Quadrivalent (Types 6, 11, 16, 18) Vaccine,
    Recombinant," updated April 30, 2009, https://wayback.archive-it
    .org/7993/20170722145339/https://www.fda.gov/BiologicsBloodVaccines
    /Vaccines/ApprovedProducts/ucm111283.htm.
5   BioSpace, "Merck & Co., Inc. Submits Biologics License Application To FDA
    For GARDASIL(R), The Company's Investigational Vaccine For Cervical
    Cancer," December 5, 2005, https://www.biospace.com/article/releases
    /merck-and-co-inc-submits-biologics-license-application-to-fda-for-gardasil
    -r-the-company-s-investigational-vaccine-for-cervical-cancer-/.
6   US Food and Drug Administration, "Prescription Drug User Fee
    Amendments," updated September 13, 2022, https://www.fda.gov/industry
    /fda-user-fee-programs/prescription-drug-user-fee-amendments.
7   US Food and Drug Administration, "Establishment of Prescription Drug
    User Fee Rates for Fiscal Year 2006," *Federal Register: The Daily Journal of
    the United States Government* 70, no. 146 (2005): 44106–44109, Document
    Number: 05-15159.
8   "Gardasil," US Food and Drug Administration, Current content as of Oct.
    24, 2019, https://www.fda.gov/vaccines-blood-biologics/vaccines/gardasil.
    *Document available to download: Supporting Documents "Approval
    History, Letters, Reviews and Related Documents-Gardasil."
9   European Medicines Agency, *PROCOMVAX: Package Insert* (West Point,
    PA: Merck & Co. Inc., 1999), https://www.ema.europa.eu/en/documents
    /product-information/procomvax-epar-product-information_en.pdf.
10 Sesilje B. Petersen and Christian Gluud, "Was Amorphous Aluminium
    Hydroxyphosphate Sulfate Adequately Evaluated Before Authorisation in

Europe?" *The BMJ: Evidence-Based Medicine* 26, no. 6 (2021): 285–289, doi:10.1136/bmjebm-2020-111419.

11 US Food and Drug Administration, *GARDASIL: Package Insert* (Whitehouse Station, NJ: Merck Sharp & Dohme Corp., a subsidiary of Merck & Co., Inc., 2006), https://www.fda.gov/files/vaccines,%20blood -%20&%20biologics/published/Package-Insert—Gardasil.pdf.

12 US Food and Drug Administration, *GARDASIL 9: Package Insert* (Whitehouse Station, NJ: Merck Sharp & Dohme Corp., a subsidiary of Merck & Co., Inc., updated 2016), https://www.immunizationinfo.com/wp-content/uploads /Gardasil-9-Prescribing-Information.pdf.

13 Ibid.

14 Trefis Team and Great Speculations, "Merck's $3 Billion Drug Jumped to 4X Growth Over Previous Year," *Forbes*, October 4, 2019, https://www.forbes .com/sites/greatspeculations/2019/10/04/mercks-3-billion-drug-jumped-to -4x-growth-over-previous-year/.

15 US Food and Drug Administration, *CERVAVIX: Package Insert* (Triangle Park, NC: GlaxoSmithKline, 2009), https://www.fda.gov/media/78013 /download.

16 Ibid.

17 Ibid.

18 Ibid.

19 US Food and Drug Administration, *GARDASIL 9: Package Insert* (Whitehouse Station, NJ: Merck Sharp & Dohme Corp., a subsidiary of Merck & Co., Inc., updated 2016), https://www.immunizationinfo.com/wp-content/uploads /Gardasil-9-Prescribing-Information.pdf.

20 Ibid.

21 Ibid.

22 Lucija Tomljenovic and Christopher A. Shaw, "Who Profits from Uncritical Acceptance of Biased Estimates of Vaccine Efficacy and Safety?," *American Journal of Public Health* 102, no. 9 (2012): e13, doi:10.2105/ AJPH.2012.300837.

23 Ibid.

24 Ibid.

25 Yukari Yaju and Hiroe Tsubaki, "Safety Concerns with Human Papilloma Virus Immunization in Japan: Analysis and Evaluation of Nagoya City's Surveillance Data for Adverse Events," *Japan Journal of Nursing Science* 16, no.4 (2019): 433–449, doi:10.1111/jjns.12252.

26 Ibid.

27 Ibid.

28 Rotem Inbar et al., "Behavioral Abnormalities in Female Mice Following Administration of Aluminum Adjuvants and the Human Papillomavirus

(HPV) Vaccine Gardasil," *Immunologic Research* 65, (2017):136–149, doi:10.1007/s12026-016-8826-6.

29 Ibid.

30 Ibid.

31 Anders Hviid et al., "Human Papillomavirus Vaccination of Adult Women and Risk of Autoimmune and Neurological Diseases," *Journal of Internal Medicine* 283, no. 2 (2018): 154–165, doi:10.1111/joim.12694.

32 David A. Geier, Janet K. Kern, and Mark R. Geier, "A Cross-Sectional Study of the Relationship Between Reported Human Papillomavirus Vaccine Exposure and the Incidence of Reported Asthma in the United States," *SAGE Open Medicine* 7, (2019): 2050312118822650, doi:10.1177/2050312118822650.

33 Ibid.

34 Lucija Tomljenovic and Christopher A. Shaw, "Who Profits from Uncritical Acceptance of Biased Estimates of Vaccine Efficacy and Safety?," *American Journal of Public Health* 102, no. 9 (2012): e13, doi:10.2105/AJPH.2012.300837.

35 Yukari Yaju and Hiroe Tsubaki, "Safety Concerns with Human Papilloma Virus Immunization in Japan: Analysis and Evaluation of Nagoya City's Surveillance Data for Adverse Events," *Japan Journal of Nursing Science* 16, no.4 (2019): 433–449, doi:10.1111/jjns.12252.

36 Rotem Inbar et al., "Behavioral Abnormalities in Female Mice Following Administration of Aluminum Adjuvants and the Human Papillomavirus (HPV) Vaccine Gardasil," *Immunologic Research* 65, (2017):136–149, doi:10.1007/s12026-016-8826-6.

37 Anders Hviid et al., "Human Papillomavirus Vaccination of Adult Women and Risk of Autoimmune and Neurological Diseases," *Journal of Internal Medicine* 283, no. 2 (2018): 154–165, doi:10.1111/joim.12694.

38 David A. Geier, Janet K. Kern, and Mark R. Geier, "A Cross-Sectional Study of the Relationship Between Reported Human Papillomavirus Vaccine Exposure and the Incidence of Reported Asthma in the United States," *SAGE Open Medicine* 7, (2019): 2050312118822650, doi:10.1177/2050312118822650.

Chapter 6

1 Lea Steele, "Prevalence and Patterns of Gulf War Illness in Kansas Veterans: Association of Symptoms with Characteristics of Person, Place, and Time of Military Service," *American Journal of Epidemiology* 152, no. 10 (2000): 992–1002, doi:10.1093/aje/152.10.992.

2 Ibid.

3 Catherine Unwin et al., "Health of UK Servicemen Who Served in Persian Gulf War," *The Lancet* 353, no. 9148 (1999): 169–178, doi:10.1016/S0140-6736(98)11338-7.

4 Ibid.

5 Ibid.

6 Matthew Hotopf et al., "Role of Vaccinations as Risk Factors for Ill Health in Veterans of the Gulf War: Cross Sectional Study," *BMJ* 320, no. 7246 (2000): 1363–1367, doi:10.1136/bmj.320.7246.1363.

7 Ibid.

8 H.L. Kelsall et al., "Symptoms and Medical Conditions in Australian Veterans of the 1991 Gulf War: Relation to Immunisations and Other Gulf War Exposures," *Occupational & Environmental Medicine* 61, no.12 (2004): 1006–1013, doi:10.1136/oem.2003.009258.

9 Ibid.

10 Lea Steele, "Prevalence and Patterns of Gulf War Illness in Kansas Veterans: Association of Symptoms with Characteristics of Person, Place, and Time of Military Service," *American Journal of Epidemiology* 152, no. 10 (2000): 992–1002, doi:10.1093/aje/152.10.992.

11 Catherine Unwin et al., "Health of UK Servicemen Who Served in Persian Gulf War," *The Lancet* 353, no. 9148 (1999): 169–178, doi:10.1016/S0140-6736(98)11338-7.

12 Matthew Hotopf et al., "Role of Vaccinations as Risk Factors for Ill Health in Veterans of the Gulf War: Cross Sectional Study," *BMJ* 320, no. 7246 (2000): 1363–1367, doi:10.1136/bmj.320.7246.1363.

13 H.L. Kelsall et al., "Symptoms and Medical Conditions in Australian Veterans of the 1991 Gulf War: Relation to Immunisations and Other Gulf War Exposures," *Occupational & Environmental Medicine* 61, no.12 (2004): 1006–1013, doi:10.1136/oem.2003.009258.

14 Lea Steele, "Prevalence and Patterns of Gulf War Illness in Kansas Veterans: Association of Symptoms with Characteristics of Person, Place, and Time of Military Service," *American Journal of Epidemiology* 152, no. 10 (2000): 992–1002, doi:10.1093/aje/152.10.992.

15 Matthew Hotopf et al., "Role of Vaccinations as Risk Factors for Ill Health in Veterans of the Gulf War: Cross Sectional Study," *BMJ* 320, no. 7246 (2000): 1363–1367, doi:10.1136/bmj.320.7246.1363.

16 Catherine Unwin et al., "Health of UK Servicemen Who Served in Persian Gulf War," *The Lancet* 353, no. 9148 (1999): 169–178, doi:10.1016/S0140-6736(98)11338-7.

17 Lea Steele, "Prevalence and Patterns of Gulf War Illness in Kansas Veterans: Association of Symptoms with Characteristics of Person, Place, and Time of

Military Service," *American Journal of Epidemiology* 152, no. 10 (2000): 992–1002, doi:10.1093/aje/152.10.992.

[18] Catherine Unwin et al., "Health of UK Servicemen Who Served in Persian Gulf War," *The Lancet* 353, no. 9148 (1999): 169–178, doi:10.1016/S0140-6736(98)11338-7.

[19] Matthew Hotopf et al., "Role of Vaccinations as Risk Factors for Ill Health in Veterans of the Gulf War: Cross Sectional Study," *BMJ* 320, no. 7246 (2000): 1363–1367, doi:10.1136/bmj.320.7246.1363.

[20] H.L. Kelsall et al., "Symptoms and Medical Conditions in Australian Veterans of the 1991 Gulf War: Relation to Immunisations and Other Gulf War Exposures," *Occupational & Environmental Medicine* 61, no.12 (2004): 1006–1013, doi:10.1136/oem.2003.009258.

Chapter 7

[1] "Who Needs a Flu Vaccine?," Centers for Disease Control and Prevention, updated September 13, 2022, https://www.cdc.gov/flu/prevent/vaccinations.htm.

[2] "Influenza (Flu) Vaccine and Pregnancy," Centers for Disease Control and Prevention, updated December 12, 2019, https://www.cdc.gov/vaccines/pregnancy/hcp-toolkit/flu-vaccine-pregnancy.html.

[3] US Food and Drug Administration, AFLURIA® QUADRIVALENT: Package Insert, STN BL 125254 (Summit, NJ: Seqirus USA Inc., 2016), https://www.fda.gov/media/117022/download.

[4] Ibid.

[5] US Food and Drug Administration, Influenza A (H1N1) 2009 Monovalent Vaccine: Package Insert, US License 1739 (Research Triangle Park, NC: GlaxoSmithKline, updated January 2010), https://www.fda.gov/media/77835/download.

[6] Elizabeth Miller et al., "Risk of Narcolepsy in Children and Young People Receiving AS03 Adjuvanted Pandemic A/H1N1 2009 Influenza Vaccine: Retrospective Analysis," *BMJ* 346 (2013): f794, doi:10.1136/bmj.f794.

[7] Ibid.

[8] Mayo Clinic, "Narcolepsy," accessed on April 13, 2023, https://www.mayoclinic.org/diseases-conditions/narcolepsy/symptoms-causes/syc-20375497.

[9] Melodie Bonvalet et al., "Autoimmunity in Narcolepsy," *Current Opinion in Pulmonary Medicine*, 23, no. 6 (2017): 522–529, doi:10.1097/MCP.0000000000000426.

[10] Ibid.

[11] Ibid.

12 Attila Szakács, Niklas Darin, and Tove Hallböök, "Increased Childhood Incidence of Narcolepsy in Western Sweden After H1N1 Influenza Vaccination," *Neurology* 80, no. 14 (2013): 1315–1321, doi:10.1212/WNL.0b013e31828ab26f.

13 Ibid.

14 Ibid.

15 Markku Partinen et al., "Increased Incidence and Clinical Picture of Childhood Narcolepsy Following the 2009 H1N1 Pandemic Vaccination Campaign in Finland," *PLoS ONE* 7, no. 3 (2012): e33723, doi:10.1371/journal.pone.0033723.

16 Ibid.

17 Carola Bardage et al., "Neurological and Autoimmune Disorders After Vaccination Against Pandemic Influenza A (H1N1) with a Monovalent Adjuvanted Vaccine: Population Based Cohort Study in Stockholm, Sweden," *BMJ* 343, (2011): d5956, doi:10.1136/bmj.d5956.

18 Ibid.

19 Jeff Kwong et al., "Risk of Guillain-Barré Syndrome after Seasonal Influenza Vaccination and Influenza Health-Care Encounters: A Self-Controlled Study," *The Lancet: Infectious Diseases* 13, no.9: 769–776, doi:10.1016/S1473-3099(13)70104-X.

20 Mayo Clinic, "Guillain-Barré syndrome," accessed on April 15, 2023, https://www.mayoclinic.org/diseases-conditions/guillain-barre-syndrome/symptoms-causes/syc-20362793

21 Jeff Kwong et al., "Risk of Guillain-Barré Syndrome after Seasonal Influenza Vaccination and Influenza Health-Care Encounters: A Self-Controlled Study," *The Lancet: Infectious Diseases* 13, no.9: 769-776, doi:10.1016/S1473-3099(13)70104-X.

22 Ibid.

23 David Juurlink et al., "Guillain-Barré Syndrome after Influenza Vaccination in Adults: A Population-Based Study," *Journal of the American Medical Association* 166, no. 20 (2006): 2217-2221, doi:10.1001/archinte.166.20.2217.

24 Jeff Kwong et al., "Risk of Guillain-Barré Syndrome After Seasonal Influenza Vaccination and Influenza Health-Care Encounters: A Self-Controlled Study," *The Lancet: Infectious Diseases* 13, no.9: 769–776, doi:10.1016/S1473-3099(13)70104-X.

25 David Juurlink et al., "Guillain-Barré Syndrome after Influenza Vaccination in Adults: A Population-Based Study," *Journal of the American Medical Association* 166, no. 20 (2006): 2217–2221, doi:10.1001/archinte.166.20.2217.

26 Ibid.

27 Tamar Lasky et al., "The Guillain–Barre Syndrome and the 1992–1993 and 1993–1994 Influenza Vaccines," *The New England Journal of Medicine* 339, no. 25 (1998): 1797–1802, doi:10.1056/NEJM199812173392501.

28 Ibid.

29 Matthew Wise et al., "Guillain-Barre Syndrome during the 2009–2010 H1N1 Influenza Vaccination Campaign: Population-Based Surveillance Among 45 Million Americans," *American Journal of Epidemiology* 175, no. 11 (2012): 1110–1119, doi:10.1093/aje/kws196.

30 Ibid.

31 Ibid.

32 Daniel A. Salmon et al., "Association Between Guillain-Barré Syndrome and Influenza A (H1N1) 2009 Monovalent Inactivated Vaccines in the USA: A Meta-Analysis," *Lancet* 381, no. 9876 (2013): 1461–1468, doi:10.1016/S0140-6736(12)62189-8.

33 Ibid.

34 Sharon Rikin et al., "Assessment of Temporally-Related Acute Respiratory Illness following Influenza Vaccination," *Vaccine* 36, no. 15 (2018): 1958–1964, doi:10.1016/j.vaccine.2018.02.105.

35 Ibid.

36 Ibid.

37 Ibid.

38 Greg G. Wolff, "Influenza Vaccination and Respiratory Virus Interference Among Department of Defense Personnel during the 2017–2018 Influenza Season," *Vaccine* 38, no. 2 (2020): 350–354, doi:10.1016/j.vaccine.2019.10.005.

39 Ibid.

40 Benjamin J. Cowling et al., "Increased Risk of Noninfluenza Respiratory Virus Infections Associated with Receipt of Inactivated Influenza Vaccine," *Clinical Infectious Diseases: An Official Publication of the Infectious Diseases Society of America* 54, no. 12 (2012): 1778–1783, doi:10.1093/cid/cis307.

41 Ibid.

42 Ibid.

43 Alexa Dierig et al., "Epidemiology of Respiratory Viral Infections in Children Enrolled in a Study of Influenza Vaccine Effectiveness," *Influenza and Other Respiratory Viruses* 8, no. 3 (2014): 293–301, doi:10.1111/irv.12229.

44 Ibid.

45 Ibid.

46 Avni Y. Joshi et al., "Effectiveness of Trivalent Inactivated Influenza Vaccine in Influenza-Related Hospitalization in Children: A Case-Control Study," *Allergy and Asthma Proceedings* 33, no. 2 (2012): e23–e27, doi:10.2500/aap.2012.33.3513.

47 Ibid.

48 Ibid.

49 Gaetano A. Lanza et al., "Inflammation-Related Effects of Adjuvant Influenza A Vaccination on Platelet Activation and Cardiac Autonomic Function," *Journal of Internal Medicine*, 269 no.1 (2011): 118–125. doi:10.1111/j.1365-2796.2010.02285.x.

50 Ibid.

51 MedAlerts.org., "Search Results From the 6/10/2022 release of VAERS data: Found 17,929 cases where Vaccine targets Influenza (FLU(H1N1) or FLU3 or FLU4 or FLUA3 or FLUA4 or FLUC3 or FLUC4 or FLUN(H1N1) or FLUN3 or FLUN4 or FLUR3 or FLUR4 or FLUX or FLUX(H1N1) or H5N1) and Standard-MedDRA-Query broadly matches 'Cardiomyopathy,'" National Vaccine Information Center, Retrieved from: https://medalerts.org/vaersdb/findfield .php? TABLE=ON&GROUP1=AGE&EVENTS=ON& SYMPTOMSSMQ=150&VAX[]=FLU(H1N1)&VAX[] =FLU3&VAX[]=FLU4&VAX[]=FLUA3&VAX[]= FLUA4&VAX[]=FLUC3&VAX[]=FLUC4&VAX[] =FLUN(H1N1)&VAX[]=FLUN3&VAX[]=FLUN4& VAX[]=FLUR3&VAX[]=FLUR4& VAX[]=FLUX&VAX[]=FLUX(H1N1)&VAX[] =H5N1&VAXTYPES=Influenza.

52 Jeff Kwong et al., "Risk of Guillain-Barré Syndrome after Seasonal Influenza Vaccination and Influenza Health-Care Encounters: A Self-Controlled Study," *The Lancet: Infectious Diseases* 13, no.9: 769–776, doi:10.1016/ S1473-3099(13)70104-X.

53 David Juurlink et al., "Guillain-Barré Syndrome after Influenza Vaccination in Adults: A Population-Based Study," *Journal of the American Medical Association* 166, no. 20 (2006): 2217–2221, doi:10.1001/archinte.166.20.2217.

54 Tamar Lasky et al., "The Guillain–Barré Syndrome and the 1992–1993 and 1993–1994 Influenza Vaccines," *The New England Journal of Medicine* 339, no. 25 (1998): 1797–1802, doi:10.1056/NEJM199812173392501.

55 Sharon Rikin et al., "Assessment of Temporally-Related Acute Respiratory Illness following Influenza Vaccination," *Vaccine* 36, no. 15 (2018): 1958–1964, doi:10.1016/j.vaccine.2018.02.105.

56 Greg G. Wolff, "Influenza Vaccination and Respiratory Virus Interference among Department of Defense Personnel During the 2017–2018 Influenza Season," *Vaccine* 38, no. 2 (2020): 350–354, doi:10.1016/j. vaccine.2019.10.005.

57 Sharon Rikin et al., "Assessment of Temporally-Related Acute Respiratory Illness following Influenza Vaccination," *Vaccine* 36, no. 15 (2018): 1958–1964, doi:10.1016/j.vaccine.2018.02.105.

58 Greg G. Wolff, "Influenza Vaccination and Respiratory Virus Interference Among Department of Defense Personnel During the 2017–2018

Influenza Season," *Vaccine* 38, no. 2 (2020): 350–354, doi:10.1016/j. vaccine.2019.10.005.

[59] Ibid.

[60] Avni Y. Joshi et al., "Effectiveness of Trivalent Inactivated Influenza Vaccine in Influenza-Related Hospitalization in Children: A Case-Control Study," *Allergy and Asthma Proceedings* 33, no. 2 (2012): e23–e27, doi:10.2500/ aap.2012.33.3513.

[61] Gaetano A. Lanza et al., "Inflammation-Related Effects of Adjuvant Influenza A Vaccination on Platelet Activation and Cardiac Autonomic Function," *Journal of Internal Medicine*, 269 no.1 (2011): 118–125. doi:10.1111/j.1365-2796.2010.02285.x.

[62] Elizabeth Miller et al., "Risk of Narcolepsy in Children and Young People Receiving AS03 Adjuvanted Pandemic A/H1N1 2009 Influenza Vaccine: Retrospective Analysis," *BMJ* 346, (2013): f794, doi:10.1136/bmj.f794.

[63] Attila Szakács, Niklas Darin, and Tove Hallböök, "Increased Childhood Incidence of Narcolepsy in Western Sweden After H1N1 Influenza Vaccination," *Neurology* 80, no. 14 (2013): 1315–1321, doi:10.1212/ WNL.0b013e31828ab26f.

[64] Markku Partinen et al., "Increased Incidence and Clinical Picture of Childhood Narcolepsy Following the 2009 H1N1 Pandemic Vaccination Campaign in Finland," *PLoS ONE* 7, no. 3 (2012): e33723, doi:10.1371/ journal.pone.0033723.

[65] Matthew Wise et al., "Guillain-Barré Syndrome during the 2009–2010 H1N1 Influenza Vaccination Campaign: Population-Based Surveillance Among 45 Million Americans," *American Journal of Epidemiology* 175, no. 11 (2012): 1110–1119, doi:10.1093/aje/kws196.

[66] Jerome I. Tokars et al., "The Risk of Guillain-Barré Syndrome Associated with Influenza A (H1N1) 2009 Monovalent Vaccine and 2009–2010 Seasonal Influenza Vaccines: Results from Self-Controlled Analyses," *Pharmacoepidemiology and Drug Safety* 21, no. 5 (2012): 546–552, doi:10.1002/pds.3220.

[67] Daniel A. Salmon et al., "Association Between Guillain-Barré Syndrome and Influenza A (H1N1) 2009 Monovalent Inactivated Vaccines in the USA: A Meta-Analysis," *Lancet* 381, no. 9876 (2013): 1461–1468, doi:10.1016/ S0140-6736(12)62189-8.

[68] Elizabeth Miller et al., "Risk of Narcolepsy in Children and Young People Receiving AS03 Adjuvanted Pandemic A/H1N1 2009 Influenza Vaccine: Retrospective Analysis," *BMJ* 346, (2013): f794, doi:10.1136/bmj.f794.

[69] Attila Szakács, Niklas Darin, and Tove Hallböök, "Increased Childhood Incidence of Narcolepsy in Western Sweden After H1N1 Influenza Vaccination," *Neurology* 80, no. 14 (2013): 1315–1321, doi:10.1212/ WNL.0b013e31828ab26f.

70 Markku Partinen et al., "Increased Incidence and Clinical Picture of Childhood Narcolepsy following the 2009 H1N1 Pandemic Vaccination Campaign in Finland," *PLoS ONE* 7, no. 3 (2012): e33723, doi:10.1371/journal.pone.0033723.

71 Carola Bardage et al., "Neurological and Autoimmune Disorders after Vaccination against Pandemic Influenza A (H1N1) with a Monovalent Adjuvanted Vaccine: Population Based Cohort Study in Stockholm, Sweden," *BMJ* 343, (2011): d5956, doi:10.1136/bmj.d5956.

72 Ibid.

73 Ibid.

74 Alexa Dierig et al., "Epidemiology of Respiratory Viral Infections in Children Enrolled in a Study of Influenza Vaccine Effectiveness," *Influenza and Other Respiratory Viruses* 8, no. 3 (2014): 293–301, doi:10.1111/irv.12229.

Chapter 8

1 Nicola P. Klein, "Licensed Pertussis Vaccines in the United States," *Human Vaccines and Immunotherapeutics* 10 no. 9: 2684–2690, doi:10.4161/hv.29576.

2 UpToDate® by Wolters Kluwer, "Diphtheria, Tetanus, and Pertussis Immunization in Children 6 weeks through 6 years of age," accessed on April 16, 2023, https://www.uptodate.com/contents/diphtheria-tetanus-and-pertussis-immunization-in-children-6-weeks-through-6-years-of-age/print.

3 Alberto Donzelli, Alessandro Schivalocchi, and Giulia Giudicatti, "Non-specific effects of vaccinations in high-income settings: How to address the issue?," *Human Vaccines & Immunotherapeutics* 14, no. 12 (2018): 2904–2910, doi:10.1080/21645515.2018.1502520.

4 Peter Aaby et al., "Evidence of Increase in Mortality after the Introduction of Diphtheria-Tetanus-Pertussis Vaccine to Children Aged 6–35 Months in Guinea-Bissau: A Time for Reflection?," *Frontiers in Public Health* 6, no. 79 (2018), doi:10.3389/fpubh.2018.00079.

5 Peter Aaby et al., "DTP with or after Measles Vaccination Is Associated with Increased In-Hospital Mortality in Guinea-Bissau," *Vaccine* 25, no. 7 (2007): 1265–1269, doi:10.1016/j.vaccine.2006.10.007.

6 "SAGE Working Group on non-specific effects of vaccines (March 2013–June 2013)," World Health Organization, accessed March 25, 2023, https://www.who.int/groups/strategic-advisory-group-of-experts-on-immunization/working-groups/non-specific-effects-of-vaccines-(march-2013—june-2013).

7 Julian P.T. Higgins et al., "Association of BCG, DTP, and Measles Containing Vaccines with Childhood Mortality: Systematic Review," *The BMJ* 355 (2016): i5170, doi:10.1136/bmj.i5170.

8   Søren Wengel Mogensen et al., "The Introduction of Diphtheria-Tetanus-Pertussis and Oral Polio Vaccine among Young Infants in an Urban African Community: A Natural Experiment," *eBioMedicine* 17 (2017): 192–198, doi:10.1016/j.ebiom.2017.01.041.

9   Ibid.

10 Peter Aaby et al., "Early Diphtheria-Tetanus-Pertussis Vaccination Associated with Higher Female Mortality and No Difference in Male Mortality in a Cohort of Low Birthweight Children: An Observational Study within a Randomised Trial," *Archives of Disease in Childhood* 97, no. 8 (2012): 685–691, doi:10.1136/archdischild-2011-300646.

11 Ibid.

12 Ibid.

13 Peter Aaby, et al., "The Introduction of Diphtheria-Tetanus-Pertussis Vaccine and Child Mortality in Rural Guinea-Bissau: An Observational Study," *International Journal of Epidemiology* 33, no. 2 (2004): 374–380, doi:10.1093/ije/dyh005.

14 Ibid.

15 Ibid.

16 Peter Aaby et al., "Is Diphtheria-Tetanus-Pertussis (DTP) Associated with Increased Female Mortality? A Meta-Analysis Testing the Hypotheses of Sex-Differential Non-Specific Effects of DTP Vaccine," *Transactions of the Royal Society of Tropical Medicine and Hygiene* 110, no. 10 (2016): 570–581, doi:10.1093/trstmh/trw073.

17 Ibid.

18 Global Advisory Committee on Vaccine Safety, 10–11 June 2004, Le; *Relevé épidémiologique hebdomadaire* 79, no. 29 (2004): 269–272, https://apps.who.int/iris/bitstream/handle/10665/232535/WER7929_269-272.PDF?sequence=1&isAllowed=y.

19 Ines Kristensen, Peter Aaby, and Henrik Jensen, "Routine Vaccinations and Child Survival: Follow Up Study in Guinea-Bissau, West Africa," *BMJ* 321, no. 7274 (2000): 1435–1438, doi:10.1136/bmj.321.7274.1435.

20 Ibid.

21 Ibid.

22 Peter Aaby et al., "Sex-Differential and Non-Specific Effects of Routine Vaccinations in a Rural Area with Low Vaccination Coverage: An Observational Study from Senegal," *Transactions of the Royal Society of Tropical Medicine and Hygiene* 109, no. 1 (2015): 77–84, doi:10.1093/trstmh/tru186.

23 Ibid.

24 Lawrence H. Moulton et al., "Evaluation of Non-Specific Effects of Infant Immunizations on Early Infant Mortality in a Southern Indian Population," *Tropical Medicine and International Health* 10, no. 10 (2005): 947–955, doi:10.1111/j.1365-3156.2005.01434.x.

25 Ibid.

26 Alexander M. Walker et al., "Diphtheria-Tetanus-Pertussis Immunization and Sudden Infant Death Syndrome," *American Journal of Public Health* 77, no. 8 (1987): 945–951, doi:10.2105/ajph.77.8.945.

27 Ibid.

28 William C. Torch, "Diphtheria-Pertussis-Tetanus (DPT) Immunization: A Potential Cause of Sudden Infant Death Syndrome," *Neurology* 32, no. 4 part 2 (1982): A169-A170.

29 Ibid.

30 Ibid.

31 Eric L. Hurwitz and Hal Morgenstern, "Effects of Diphtheria-Tetanus Pertussis or Tetanus Vaccination on Allergies and Allergy-Related Respiratory Symptoms Among Children and Adolescents in the United States," *Journal of Manipulative and Physiological Therapeutics* 23, no. 2 (2000): 81–90, doi:10.1016/S0161-4754(00)90072-1.

32 Ibid.

33 Ibid.

34 Kara L. McDonald et al., "Delay in Diphtheria, Pertussis, Tetanus Vaccination Is Associated with a Reduced Risk of Childhood Asthma," *The Journal of Allergy and Clinical Immunology* 121, no. 3 (2008): 626–631, doi:10.1016/j.jaci.2007.11.034.

35 Ibid.

36 Ibid.

37 Ibid.

38 Tricia M. McKeever et al. "Vaccination and Allergic Disease: A Birth Cohort Study," *American Journal of Public Health* 94 (2004) 985–989, doi:10.2105/ajph.94.6.985.

39 Ibid.

40 Ibid.

41 Jason M. Glanz et al., "A Population-Based Cohort Study of Undervaccination in 8 Managed Care Organizations across the United States," *JAMA Pediatrics* 167 no. 1 (2013): 274–281, doi: 10.1001/jamapediatrics.2013.502.

42 Ibid.

43 Tricia M. McKeever et al., "Vaccination and Allergic Disease: A Birth Cohort Study," *American Journal of Public Health* 94 (2004) 985–989, doi:10.2105/ajph.94.6.985.

44 Søren Wengel Mogensen et al., "The Introduction of Diphtheria-Tetanus-Pertussis and Oral Polio Vaccine Among Young Infants in an Urban African Community: A Natural Experiment," *eBioMedicine* 17 (2017): 192–198, doi:10.1016/j.ebiom.2017.01.041.

45 Peter Aaby et al., "Early Diphtheria-Tetanus-Pertussis Vaccination Associated with Higher Female Mortality and No Difference in Male Mortality in a Cohort

of Low Birthweight Children: An Observational Study within a Randomised Trial," *Archives of Disease in Childhood* 97, no. 8 (2012): 685–691, doi:10.1136 /archdischild-2011-300646.

46 Peter Aaby, et al., "The Introduction of Diphtheria-Tetanus-Pertussis Vaccine and Child Mortality in Rural Guinea-Bissau: An Observational Study," *International Journal of Epidemiology* 33, no. 2 (2004): 374–380, doi:10.1093/ ije/dyh005.

47 Peter Aaby et al., "Is Diphtheria-Tetanus-Pertussis (DTP) Associated with Increased Female Mortality? A Meta-Analysis Testing the Hypotheses of Sex-Differential Non-Specific Effects of DTP Vaccine," *Transactions of the Royal Society of Tropical Medicine and Hygiene* 110, no. 10 (2016): 570–581, doi:10.1093/trstmh/trw073.

48 Ines Kristensen, Peter Aaby, and Henrik Jensen, "Routine Vaccinations and Child Survival: Follow-Up Study in Guinea-Bissau, West Africa," *BMJ* 321, no. 7274 (2000): 1435–1438, doi:10.1136/bmj.321.7274.1435.

49 Peter Aaby et al., "Sex-Differential and Non-Specific Effects of Routine Vaccinations in a Rural Area with Low Vaccination Coverage: An Observational Study from Senegal," *Transactions of the Royal Society of Tropical Medicine and Hygiene* 109, no. 1 (2015): 77–84, doi:10.1093/trstmh/tru186.

50 Lawrence H. Moulton et al., "Evaluation of Non-Specific Effects of Infant Immunizations on Early Infant Mortality in a Southern Indian Population," *Tropical Medicine and International Health* 10, no. 10 (2005): 947–955, doi:10.1111/j.1365-3156.2005.01434.x.

51 Ines Kristensen, Peter Aaby, and Henrik Jensen, "Routine Vaccinations and Child Survival: Follow-Up Study in Guinea-Bissau, West Africa," *BMJ* 321, no. 7274 (2000): 1435–1438, doi:10.1136/bmj.321.7274.1435.

52 Peter Aaby et al., "Is Diphtheria-Tetanus-Pertussis (DTP) Associated with Increased Female Mortality? A Meta-Analysis Testing the Hypotheses of Sex-Differential Non-Specific Effects of DTP Vaccine," *Transactions of the Royal Society of Tropical Medicine and Hygiene* 110, no. 10 (2016): 570–581, doi:10.1093/trstmh/trw073.

53 Peter Aaby et al., "The Introduction of Diphtheria-Tetanus-Pertussis Vaccine and Child Mortality in Rural Guinea-Bissau: An Observational Study," *International Journal of Epidemiology* 33, no. 2 (2004): 374–380, doi:10.1093/ ije/dyh005.

54 Peter Aaby et al., "Early Diphtheria-Tetanus-Pertussis Vaccination Associated with Higher Female Mortality and No Difference in Male Mortality in a Cohort of Low Birthweight Children: An Observational Study within a Randomised Trial," *Archives of Disease in Childhood* 97, no. 8 (2012): 685–691, doi:10.1136 /archdischild-2011-300646.

55 Søren Wengel Mogensen et al., "The Introduction of Diphtheria-Tetanus-Pertussis and Oral Polio Vaccine among Young Infants in an Urban African Community: A Natural Experiment," *eBioMedicine* 17 (2017): 192–198, doi:10.1016/j.ebiom.2017.01.041.

56 Peter Aaby et al., "DTP With or After Measles Vaccination is Associated with Increased In-Hospital Mortality in Guinea-Bissau," *Vaccine* 25, no. 7 (2007): 1265–1269, doi:10.1016/j.vaccine.2006.10.007.

57 Peter Aaby et al., "Sex-Differential and Non-Specific Effects of Routine Vaccinations in a Rural Area with Low Vaccination Coverage: An Observational Study from Senegal," *Transactions of the Royal Society of Tropical Medicine and Hygiene* 109, no. 1 (2015): 77 84, doi:10.1093/trstmh/tru186.

58 Alexander M. Walker et al., "Diphtheria-Tetanus-Pertussis Immunization and Sudden Infant Death Syndrome," *American Journal of Public Health* 77, no. 8 (1987): 945–951, doi:10.2105/ajph.77.8.945.

59 William C. Torch, "Diphtheria-Pertussis-Tetanus (DPT) Immunization: A Potential Cause of Sudden Infant Death Syndrome," *Neurology* 32, no. 4 part 2 (1982): A169-A170.

60 Eric L. Hurwitz and Hal Morgenstern, "Effects of Diphtheria-Tetanus-Pertussis or Tetanus Vaccination on Allergies and Allergy-Related Respiratory Symptoms Among Children and Adolescents in the United States," *Journal of Manipulative and Physiological Therapeutics* 23, no. 2 (2000): 81–90, doi:10.1016/S0161-4754(00)90072-1.

61 Kara L. McDonald et al., "Delay in Diphtheria, Pertussis, Tetanus Vaccination Is Associated with a Reduced Risk of Childhood Asthma," *The Journal of Allergy and Clinical Immunology* 121, no. 3 (2008): 626–631, doi:10.1016/j.jaci.2007.11.034.

62 Tricia M. McKeever et al. "Vaccination and Allergic Disease: A Birth Cohort Study," *American Journal of Public Health* 94 (2004) 985–989, doi:10.2105/ajph.94.6.985.

63 Ibid.

Chapter 9

1 "Hepatitis B Vaccination of Infants, Children, and Adolescents," U.S. Centers for Disease Control, accessed March 26, 2023, https://www.cdc.gov/hepatitis/hbv/vaccchildren.htm.

2 Monica A. Fisher and Stephen A. Eklund, "Hepatitis B Vaccine and Liver Problems in U.S. Children Less than 6 Years Old, 1993 and 1994," *Epidemiology* 10, no. 3 (1999): 337–339, https://journals.lww.com/epidem/Abstract/1999/05000/Hepatitis_B_Vaccine_and_Liver_Problems_in_U_S_.24.aspx.

3   Ibid.

4   Ibid.

5   Ibid.

6   Nancy Agmon-Levin et al., "Immunization with Hepatitis B Vaccine Accelerates SLE-Like Disease in a Murine Model," *Journal of Autoimmunity* 54, (2014): 21–32, doi:10.1016/j.jaut.2014.06.006.

7   Ibid.

8   Ibid.

9   David C. Classen and John Barthelow Classen, "The Timing of Pediatric Immunization and the Risk of Insulin-Dependent Diabetes Mellitus," *Infectious Diseases in Clinical Practice* 6, no. 7 (1997): 449–454, https://journals.lww.com/infectdis/citation/1997/06070/the_timing_of_pediatric_immunization_and_the_risk.7.aspx.

10 Ibid.

11 Ibid.

12 Miguel A. Hernán et al., "Recombinant Hepatitis B Vaccine and the Risk of Multiple Sclerosis: A Prospective Study," *Neurology* 63, no. 5 (2004): 838–842, doi:10.1212/01.wnl.0000138433.61870.82.

13 Ibid.

14 Ibid.

15 Dong Keon Yon et al., "Hepatitis B Immunogenicity After a Primary Vaccination Course Associated with Childhood Asthma, Allergic Rhinitis, and Allergen Sensitization." *Pediatric Allergy and Immunology: Official Publication of the European Society of Pediatric Allergy and Immunology* 29, no. 2 (2018): 221–224, doi:10.1111/pai.12850.

16 Ibid.

17 Ibid.

18 Ibid.

19 "VAERS Data," VAERS, accessed September 23, 2022, https://vaers.hhs.gov/data.html.

20 Ibid.

21 Penina Haber et al., "Safety of Currently Licensed Hepatitis B Surface Antigen Vaccines in the United States, Vaccine Adverse Event Reporting System (VAERS), 2005–2015," *Vaccine* 36, no. 4 (2018): 559–564, doi:10.1016/j.vaccine.2017.11.079.

22 Ibid.

23 Ibid.

24 Ibid.

25 Young June Choe et al., "Sudden Death in the First 2 Years of Life following Immunization in the Republic of Korea," *Pediatrics international: Official*

Journal of the Japan Pediatric Society 54, no.6 (2012): 905–910, doi:10.1111/j.1442-200X.2012.03697.x.

26 Monica A. Fisher and Stephen A. Eklund, "Hepatitis B Vaccine and Liver Problems in U.S. Children Less than 6 Years Old, 1993 and 1994," *Epidemiology* 10, no. 3 (1999): 337–339, https://journals.lww.com/epidem/Abstract/1999/05000/Hepatitis_B_Vaccine_and_Liver_Problems_in_U_S_.24.aspx

27 Nancy Agmon-Levin et al., "Immunization with Hepatitis B Vaccine Accelerates SLE-Like Disease in a Murine Model," *Journal of Autoimmunity* 54, (2014): 21–32, doi:10.1016/j.jaut.2014.06.006.

28 David C. Classen and John Barthelow Classen, "The Timing of Pediatric Immunization and the Risk of Insulin-Dependent Diabetes Mellitus," *Infectious Diseases in Clinical Practice* 6, no. 7 (1997): 449–454, https://journals.lww.com/infectdis/citation/1997/06070/the_timing_of_pediatric_immunization_and_the_risk.7.aspx.

29 Miguel A. Hernán et al., "Recombinant Hepatitis B Vaccine and the Risk of Multiple Sclerosis: A Prospective Study," *Neurology* 63, no. 5 (2004): 838–842, doi:10.1212/01.wnl.0000138433.61870.82.

30 Dong Keon Yon et al., "Hepatitis B Immunogenicity after a Primary Vaccination Course Associated with Childhood Asthma, Allergic Rhinitis, and Allergen Sensitization." *Pediatric Allergy and Immunology: Official Publication of the European Society of Pediatric Allergy and Immunology* 29, no. 2 (2018): 221–224, doi:10.1111/pai.12850.

31 "VAERS Data," VAERS, accessed September 23, 2022, https://vaers.hhs.gov/data.html.

32 Nancy Agmon-Levin et al., "Immunization with Hepatitis B Vaccine Accelerates SLE-Like Disease in a Murine Model," *Journal of Autoimmunity* 54, (2014): 21–32, doi:10.1016/j.jaut.2014.06.006.

Chapter 10

1 Kenichiro Sato et al., "Facial Nerve Palsy Following the Administration of COVID-19 mRNA Vaccines: Analysis of a Self-Reporting Database," *International Journal of Infectious Diseases : IJID : Official Publication of the International Society for Infectious Diseases* 111, (2021): 310–312, doi:10.1016/j.ijid.2021.08.071.

2 Ibid.

3 National Institute of Neurological Disorders and Stroke, "Bell's Palsy," accessed on April 16, 2023, https://www.ninds.nih.gov/health-information/disorders/bells-palsy.

4   Erik Y. F. Wan et al., "Bell's Palsy Following Vaccination with mRNA (BNT162b2) and Inactivated (CoronaVac) SARS-CoV-2 Vaccines: A Case Series and Nested Case-Control Study," *The Lancet Infectious Diseases* 22, no. 1 (2022): 64–72, doi:10.1016/S1473-3099(21)00451-5.

5   Ibid.

6   Rana Shibili et al., "Association Between Vaccination with the BNT162b2 mRNA COVID-19 Vaccine and Bell's Palsy: A Population-Based Study," *The Lancet Regional Health. Europe* 11 (2021); 100236, doi:10.1016/j.lanepe.2021.100236.

7   Ibid.

8   Ibid.

9   Ibid.

10 Eric Yuk Fai Wan et al., "Messenger RNA Coronavirus Disease 2019 (COVID-19) Vaccination With BNT162b2 Increased Risk of Bell's Palsy: A Nested Case-Control and Self-Controlled Case Series Study," Clinical Infectious Diseases: An Official Publication of the Infectious Diseases Society of America 76, no. 3 (2023); e291–e298, doi:10.1093/cid/ciac460.

11 Ibid.

12 Ibid.

13 Min S. Kim et al., "Comparative Safety of mRNA COVID-19 Vaccines to Influenza Vaccines: A Pharmacovigilance Analysis Using WHO International Database," *Journal of Medical Virology* 94, no. 3 (2022), doi:10.1002/jmv.27424.

14 Ibid.

15 Francisco T. T. Lai et al., "Adverse Events of Special Interest Following the Use of BNT162b2 in Adolescents: A Population-Based Retrospective Cohort Study," *Emerging Microbes and Infections* 11, no.1 (2022): 885–893, doi:10.1080/22221751.2022.2050952.

16 Ibid.

17 Ibid.

18 Ibid.

19 Cleveland Clinic, "Myocarditis," accessed on April 16, 2023, https://my.clevelandclinic.org/health/diseases/22129-myocarditis.

20 Ibid.

21 Øystein Karlstad et al., "SARS-CoV-2 Vaccination and Myocarditis in a Nordic Cohort Study of 23 Million Residents," *Journal of American Medical Association Cardiology* 7, no. 6 (2022): 600–612, doi:10.1001/jamacardio.2022.0583.

22 Ibid.

23 Martina Patone et al., "Risk of Myocarditis After Sequential Doses of COVID-19 Vaccine and SARS-CoV-2 Infection by Age and Sex," *Circulation* 146, no. 10: 743–754, doi:10.1161/CIRCULATIONAHA.122.059970.

24 Ibid.

25 Anthony Simone et al., "Acute Myocarditis Following a Third Dose of COVID-19 mRNA Vaccination in Adults," *International Journal of Cardiology, 365* (2022): 41–43, doi:10.1016/j.ijcard.2022.07.031.

26 Ibid.

27 Ibid.

28 Francisco Tsz Tsun Lai et al., "Carditis After COVID-19 Vaccination With a Messenger RNA Vaccine and an Inactivated Virus Vaccine: A Case-Control Study," *Annals of Internal Medicine* 175, no. 3 (2022); 362–370, doi:10.7326/M21-3700.

29 Ibid.

30 Ibid.

31 Ibid.

32 Dror Mevorach et al., "Myocarditis after BNT162b2 mRNA Vaccine against Covid-19 in Israel," *The New England Journal of Medicine* 385, no. 23 (2021); 2140–2149, doi:10.1056/NEJMoa2109730.

33 Ibid.

34 Ibid.

35 Marco Massari et al., "Postmarketing Active Surveillance of Myocarditis and Pericarditis Following Vaccination with COVID-19 mRNA Vaccines in Persons Aged 12 to 39 years in Italy: A Multi-Database, Self-Controlled Case Series Study," *PLoS Medicine* 19, no. 7 (2022): e1004056, doi:10.1371/journal.pmed.1004056.

36 Ibid.

37 Kristin Goddard et al., "Risk of Myocarditis and Pericarditis following BNT162b2 and mRNA-1273 COVID-19 Vaccination," *Vaccine* 40, no. 35 (2022): 5153–5159, doi:10.1016/j.vaccine.2022.07.007.

38 Ibid.

39 C.R. Simpson et al., "First-Dose ChAdOx1 and BNT162b2 COVID-19 Vaccines and Thrombocytopenic, Thromboembolic and Hemorrhagic Events in Scotland," *Nature Medicine* 27, no. 7 (2021); 1290–1297, doi:10.1038/s41591-021-01408-4.

40 Ibid.

41 Ibid.

42 Jacob D. Berild et al., "Analysis of Thromboembolic and Thrombocytopenic Events After the AZD1222, BNT162b2, and MRNA-1273 COVID-19 Vaccines in 3 Nordic Countries," *Journal of the American Medical Association Network Open* 5, no. 6: e2217375, doi:10.1001/jamanetworkopen.2022.17375.

43 Ibid.

44 Erik Y.F. Wan et al., "Herpes Zoster Related Hospitalization after Inactivated (CoronaVac) and mRNA (BNT162b2) SARS-CoV-2 Vaccination: A Self-Controlled Case Series and Nested Case-Control Study," *The Lancet Regional Health: Western Pacific* 21, no. 100393 (2022), doi:10.1016/j.lanwpc.2022.100393.

45 Ibid.

46 Ibid.

47 Yoav Yanir et al., "Association Between the BNT162b2 Messenger RNA COVID-19 Vaccine and the Risk of Sudden Sensorineural Hearing Loss," *Journal of the American Medical Association–Otolaryngology—Head and Neck Surgery* 148, no. 4 (2022): 299–306, doi:10.1001/jamaoto.2021.4278.

48 Ibid.

49 Diego Montano, "Frequency and Associations of Adverse Reactions of COVID-19 Vaccines Reported to Pharmacovigilance Systems in the European Union and the United States," *Frontiers in Public Health* 9 (2022): 756633, doi:10.3389/fpubh.2021.756633.

50 Ibid.

51 Ibid.

52 Hui-Lee Wong et al., "Surveillance of COVID-19 Vaccine Safety among Elderly Persons Aged 65 Years and Older," *Vaccine* 41, no. 2 (2023): 532–539, doi:10.1016/j.vaccine.2022.11.069.

53 Ibid.

54 Joseph Fraiman et al., "Serious Adverse Events of Special Interest following mRNA COVID-19 Vaccination in Randomized Trials in Adults," *Vaccine* 40, no. 40 (2022): 5798–5805, doi:10.1016/j.vaccine.2022.08.036.

55 Ibid.

56 Ibid.

57 Ibid.

58 Kristin Goddard et al., Risk of Myocarditis and Pericarditis Following BNT162b2 and mRNA-1273 COVID-19 Vaccination," *Vaccine* 40, no. 35 (2022): 5153–5159, doi:10.1016/j.vaccine.2022.07.007.

59 Francisco T. T. Lai et al., "Adverse Events of Special Interest Following the Use of BNT162b2 in Adolescents: A Population-Based Retrospective Cohort Study," *Emerging Microbes and Infections* 11, no.1 (2022): 885–893, doi:10.1080/22221751.2022.2050952.

60 Marco Massari et al., "Postmarketing Active Surveillance of Myocarditis and Pericarditis following Vaccination with COVID-19 mRNA Vaccines in Persons Aged 12 to 39 years in Italy: A Multi-Database, Self-Controlled Case Series Study," *PLoS Medicine* 19, no. 7 (2022): e1004056, doi:10.1371/journal.pmed.1004056.

61 Anthony Simone et al., "Acute Myocarditis Following a Third Dose of COVID-19 mRNA Vaccination in Adults," *International Journal of Cardiology, 365* (2022): 41–43, doi:10.1016/j.ijcard.2022.07.031.

62 Øystein Karlstad et al., "SARS-CoV-2 Vaccination and Myocarditis in a Nordic Cohort Study of 23 Million Residents," *Journal of American Medical Association Cardiology* 7, no. 6 (2022): 600–612, doi:10.1001/jamacardio.2022.0583.

63 Martina Patone et al., "Risk of Myocarditis After Sequential Doses of COVID-19 Vaccine and SARS-CoV-2 Infection by Age and Sex," *Circulation* 146, no. 10: 743–754, doi:10.1161/CIRCULATIONAHA.122.059970.

64 Hui Lee Wong et al., "Surveillance of COVID-19 Vaccine Safety Among Elderly Persons Aged 65 Years and Older," *Vaccine* 41, no. 2 (2023): 532–539, doi:10.1016/j.vaccine.2022.11.069.

65 Dror Mevorach et al., "Myocarditis after BNT162b2 mRNA Vaccine against Covid-19 in Israel," *The New England Journal of Medicine* 385, no. 23 (2021); 2140–2149, doi:10.1056/NEJMoa2109730.

66 Eric Yuk Fai Wan et al., "Messenger RNA Coronavirus Disease 2019 (COVID-19) Vaccination with BNT162b2 Increased Risk of Bell's Palsy: A Nested Case-Control and Self-Controlled Case Series Study," Clinical Infectious Diseases: An Official Publication of the Infectious Diseases Society of America 76, no. 3 (2023); e291–e298, doi:10.1093/cid/ciac460.

67 Kenichiro Sato et al., "Facial Nerve Palsy following the Administration of COVID-19 mRNA Vaccines: Analysis of a Self-Reporting Database," *International Journal of Infectious Diseases : IJID : Official Publication of the International Society for Infectious Diseases* 111, (2021): 310–312, doi:10.1016/j.ijid.2021.08.071.

68 Rana Shibili et al., "Association between Vaccination with the BNT162b2 mRNA COVID-19 Vaccine and Bell's Palsy: A Population-Based Study," *The Lancet Regional Health. Europe* 11 (2021); 100236, doi:10.1016/j.lanepe.2021.100236.

69 Erik Y.F. Wan et al., "Herpes Zoster Related Hospitalization after Inactivated (CoronaVac) and mRNA (BNT162b2) SARS-CoV-2 Vaccination: A Self-Controlled Case Series and Nested Case-Control Study," *The Lancet Regional Health: Western Pacific* 21, no. 100393 (2022), doi:10.1016/j.lanwpc.2022.100393.

Chapter 11

1 Medicines Adverse Reactions Committee, "Use of Boostrix (Combined Diphtheria, Tetanus and Pertussis Vaccine) in Pregnancy: Confidential," report (2020), https://www.medsafe.govt.nz/committees/marc/reports/181-Use-of-Boostrix.pdf.

2 US Food and Drug Administration, *Fluvirin®: Package Insert*, (Summit, NJ: Seqirus USA Inc., Revised 2017), https://www.fda.gov/files/vaccines%2C%20 -blood%20%26%20biologics/published/Package-Insert—Fluvirin.pdf.

3 US Food and Drug Administration, *Comirnaty®: Package Insert* (New York, NY: Pfizer Inc., 2022), https://www.fda.gov/media/151707/download.

4 US Food and Drug Administration, *Spikevax®: Package Insert* (New York, NY: Moderna Inc., 2022), https://www.fda.gov/media/155675/download.

5 "Pregnancy Guidelines and Recommendations by Vaccine," Centers for Disease Control and Prevention, August 31, 2016, https://www.cdc.gov /vaccines/pregnancy/hcp-toolkit/guidelines.html.

6 "Covid-19 Vaccines While Pregnant or Breastfeeding," Centers for Disease Control and Prevention, Updated June 16, 2022, https://www.cdc.gov /coronavirus/2019-ncov/vaccines/recommendations/pregnancy.html.

7 Centers for Disease Control and Prevention (2021), "COVID-19 Vaccine Pregnancy Registry," Vaccine Safety, accessed May 3, 2023. https://www.cdc .gov/vaccinesafety/ensuringsafety/monitoring/v-safe/covid-preg-reg.html.

8 Ousseny Zerbo et al., "Association between Influenza Infection and Vaccination During Pregnancy and Risk of Autism Spectrum Disorder," *JAMA Pediatrics* 171, no. 1 (2017): e163609, doi:10.1001/jamapediatrics.2016.3609.

9 Ibid.

10 Ibid.

11 Juliet Popper Shaffer, "Multiple Hypothesis Testing," *Annual Review of Psychology* 46, (1995): 561–584, http://wexler.free.fr/library/files/shaffer%20 -(1995)%20multiple%20hypothesis%20testing.pdf.

12 Alberto Donzelli, Alessandro Schivalocchi, and Alessandro Battaggia, "Influenza Vaccination in the First Trimester of Pregnancy and Risk of Autism Spectrum Disorder," *JAMA Pediatrics* 171, (2017): 601, doi:10.1001/ jamapediatrics.2017.0753.

13 Brian S. Hooker, "Influenza Vaccination in the First Trimester of Pregnancy and Risk of Autism Spectrum Disorder," *JAMA Pediatrics* 171, no. 6 (2007): 600, doi:10.1001/jamapediatrics.2017.0734.

14 Ousseny Zerbo et al., "Association between Influenza Infection and Vaccination During Pregnancy and Risk of Autism Spectrum Disorder," *JAMA Pediatrics* 171, no. 1 (2017): e163609, doi:10.1001/jamapediatrics.2016.3609.

15 Ibid.

16 Brian S. Hooker, "Influenza Vaccination in the First Trimester of Pregnancy and Risk of Autism Spectrum Disorder," *JAMA Pediatrics* 171, no. 6 (2007): 600, doi:10.1001/jamapediatrics.2017.0734.

17 Stephanie A. Irving et al., "Trivalent Inactivated Influenza Vaccine and Spontaneous Abortion," *Obstetrics and Gynecology* 121, no. 1 (2013): 159–165, doi:10.1097/aog.0b013e318279f56f.

18 Ibid.

19 Ibid.

20 James G. Donahue et al., "Association of Spontaneous Abortion with Receipt of Inactivated Influenza Vaccine Containing H1N1pdm09 in 2010-11 and 2011-12," *Vaccine* 35, no. 40 (2017): 5314–5322, doi:10.1016/j.vaccine.2017.06.069.

21 James G. Donahue et al., "Inactivated Influenza Vaccine and Spontaneous Abortion in the Vaccine Safety Datalink in 2012–13, 2013–14, and 2014–15," *Vaccine* 37, no.44 (2019): 6673–6681, doi:10.1016/j.vaccine.2019.09.035.

22 Stephanie A. Irving et al., "Trivalent Inactivated Influenza Vaccine and Spontaneous Abortion," *Obstetrics and Gynecology* 121, no. 1 (2013): 159–165, doi:10.1097/aog.0b013e318279f56f.

23 Gary S. Goldman, "Comparison of VAERS Fetal-Loss Reports during Three Consecutive Influenza Seasons: Was There a Synergistic Fetal Toxicity Associated with the Two-Vaccine 2009/2010 Season?," *Human & Experimental Toxicology 32, no. 5* (2012) 464–475. https://doi.org/10.1177/0960327112455067.

24 Ibid.

25 Ibid.

26 Ibid.

27 Ibid.

28 Ibid.

29 James G. Donahue et al., "Association of Spontaneous Abortion with Receipt of Inactivated Influenza Vaccine Containing H1N1pdm09 in 2010–11 and 2011–12," *Vaccine* 35, no. 40 (2017): 5314–5322, doi:10.1016/j.vaccine.2017.06.069.

30 Ibid.

31 Ibid.

32 James G. Donahue et al., "Inactivated Influenza Vaccine and Spontaneous Abortion in the Vaccine Safety Datalink in 2012–13, 2013–14, and 2014–15," *Vaccine* 37 (2019): 6673–6681, doi:10.1016/j.vaccine.2019.09.035.

33 Ibid.

34 Ibid.

35 Alberto Donzelli, "Influenza Vaccination of Pregnant Women and Serious Adverse Events in the Offspring," *International Journal of Environmental Research and Public Health* 16, no. 22 (2019): 4347, doi:10.3390/ijerph16224347.

36 Milagritos Tapia et al., "Maternal Immunisation with Trivalent Inactivated Influenza Vaccine for Prevention of Influenza in Infants in Mali: A Prospective, Active-controlled, Observer-blind, Randomised Phase 4 Trial,"

The Lancet. Infectious Diseases 16, no. 9 (2016): 1026–1035. doi:10.1016/S1473-3099(16)30054-8.

37   Alberto Donzelli, "Influenza Vaccination of Pregnant Women and Serious Adverse Events in the Offspring," *International Journal of Environmental Research and Public Health* 16, no. 22 (2019): 4347, doi:10.3390/ijerph16224347.

38   Ibid.

39   Ibid.

40   Milagritos Tapia et al., "Maternal Immunisation with Trivalent Inactivated Influenza Vaccine for Prevention of Influenza in Infants in Mali: A Prospective, Active-controlled, Observer-blind, Randomised Phase 4 Trial," *The Lancet. Infectious Diseases* 16, no. 9 (2016): 1026–1035. doi:10.1016/S1473-3099(16)30054-8.

41   Ibid.

42   Alberto Donzelli, "Influenza Vaccination for All Pregnant Women? So Far the Less Biased Evidence does not Favour It," *Human Vaccines and Immunotherapeutics* 15, no. 9 (2019): 2159–2164, doi:10.1080/21645515.2019.1568161.

43   Ibid.

44   Ibid.

45   Lisa M. Christian et al., "Inflammatory Responses to Trivalent Influenza Virus Vaccine among Pregnant Women," *Vaccine* 29, no. 48, (2011): 8982–8987, doi:10.1016/j.vaccine.2011.09.039.

46   Ibid.

47   Ibid.

48   Ibid.

49   Ibid.

50   Cristopher S. Price et al., "Prenatal and Infant Exposure to Thimerosal from Vaccines and Immunoglobulins and Risk of Autism," *Pediatrics* 126, no. 4 (2010): 656–664, doi:10.1542/peds.2010-0309.

51   Ibid.

52   Ibid.

53   US Food and Drug Administration, *Fluvirin®: Package Insert*, (Summit, NJ: Seqirus USA Inc., Revised 2017), https://www.fda.gov/files/vaccines%2C%20-blood%20%26%20biologics/published/Package-Insert—Fluvirin.pdf.

54   Cristopher S. Price et al., "Prenatal and Infant Exposure to Thimerosal from Vaccines and Immunoglobulins and Risk of Autism," *Pediatrics* 126, no. 4 (2010): 656–664, doi:10.1542/peds.2010-0309.

55   Cristopher S. Price, Anne Robertson, and Barbara Goodson, "Thimerosal and Autism Technical Report," Abt Associates 1, (2009): https://www.abtassociates.com/insights/publications/report/thimerosal-and-autism-technical-report-volume-1.

56 Ibid.

57 Elyse O. Kharbanda et al., "Evaluation of the Association of Maternal Pertussis Vaccination with Obstetric Events and Birth Outcomes," *JAMA* 312, no. 18 (2014): 1897–1904, doi:10.1001/jama.2014.14825.

58 Ibid.

59 Ibid.

60 Ibid.

61 J.B. Layton et al., "Prenatal Tdap Immunization and Risk of Maternal and Newborn Adverse Events," *Vaccine* 35, no. 33 (2017): 4072–4078, doi:10.1016/j.vaccine.2017.06.071.

62 Ibid.

63 Ibid.

64 Ibid.

65 Ibid.

66 Ibid.

67 Malini DeSilva et al., "Maternal Tdap Vaccination and Risk of Infant Morbidity," Vaccine 35, no. 29 (2017): 3655–3660, doi:10.1016/j.vaccine.2017.05.041.

68 Ibid.

69 Pedro Moro et al., "Enhanced Surveillance of Tetanus Toxoid, Reduced Diphtheria Toxoid, and Acellular Pertussis (Tdap) Vaccines in Pregnancy in the Vaccine Adverse Event Reporting System (VAERS), 2011-2015," *Vaccine* 34, no. 20 (2016): 2349–2353, doi:10.1016/j.vaccine.2016.03.049.

70 Ibid.

71 Ibid.

72 Malini DeSilva et al., "Evaluation of Acute Adverse Events after Covid-19 Vaccination during Pregnancy," *The New England Journal of Medicine* 387, no. 2 (2022): 187–189, doi:10.1056/NEJMc2205276.

73 US Food and Drug Administration, *Comirnaty®: Package Insert,* (New York, NY: Pfizer Inc., 2021), https://www.fda.gov/media/154834/download.

74 "Covid-19 Vaccines While Pregnant or Breastfeeding," Centers for Disease Control and Prevention, updated October 20, 2022, https://www.cdc.gov/coronavirus/2019-ncov/vaccines/recommendations/pregnancy.html.

75 Malini DeSilva et al., "Evaluation of Acute Adverse Events after Covid-19 Vaccination during Pregnancy," *The New England Journal of Medicine* 387, no. 2 (2022): 187–189, doi:10.1056/NEJMc2205276.

76 Aharon Dick et al., "Safety of Third SARS-CoV-2 Vaccine (Booster Dose) During Pregnancy," *American Journal of Obstetrics & Gynecology MFM 4*, no. 4 (2022): 100637, doi:10.1016/j.ajogmf.2022.100637.

77 Ibid.

78 Ibid.

79 "Gestational Diabetes," Centers for Disease Control and Prevention, accessed on April 16, 2023, https://www.cdc.gov/diabetes/basics/gestational.html.

80 "VAERS Data," Vaccine Adverse Event Reporting System (VAERS), updated April 7, 2023, https://vaers.hhs.gov/data.html.

81 Ibid.

82 Ibid.

83 Itai Gat et al., "Covid-19 Vaccination BNT162b2 Temporarily Impairs Semen Concentration and Total Motile Count among Semen Donors," *Andrology* 10, no. 6 (2022): 1016–1022, doi:10.1111/andr.13209.

84 Ibid.

85 Ibid.

86 Ibid.

87 Stephanie A. Irving et al., "Trivalent Inactivated Influenza Vaccine and Spontaneous Abortion," *Obstetrics and Gynecology* 121, no. 1 (2013): 159–165, doi:10.1097/aog.0b013e318279f56f.

88 Gary S. Goldman, "Comparison of VAERS Fetal-Loss Reports during Three Consecutive Influenza Seasons: Was There a Synergistic Fetal Toxicity Associated with the Two-Vaccine 2009/2010 Season?," *Human & Experimental Toxicology 32, no.* 5 (2012) 464–475, https://doi.org/10.1177/0960327112455067.

89 James G. Donahue et al., "Inactivated Influenza Vaccine and Spontaneous Abortion in the Vaccine Safety Datalink in 2012–13, 2013–14, and 2014–15," *Vaccine* 37, no. 44 (2019): 6673–6681, doi:10.1016/j.vaccine.2019.09.035.

90 Alberto Donzelli, "Influenza Vaccination of Pregnant Women and Serious Adverse Events in the Offspring," *International Journal of Environmental Research and Public Health* 16, no. 22 (2019): 4347, doi:10.3390/ijerph16224347.

91 Ousseny Zerbo et al., "Association between Influenza Infection and Vaccination During Pregnancy and Risk of Autism Spectrum Disorder," *JAMA Pediatrics* 171, no. 1 (2017): e163609, doi:10.1001/jamapediatrics.2016.3609.

92 Cristopher S. Price et al., "Prenatal and Infant Exposure to Thimerosal from Vaccines and Immunoglobulins and Risk of Autism," *Pediatrics* 126, no. 4 (2010): 656–664, doi:10.1542/peds.2010-0309.

93 Alberto Donzelli, "Influenza Vaccination of Pregnant Women and Serious Adverse Events in the Offspring," *International Journal of Environmental Research and Public Health* 16, no. 22 (2019): 4347, doi:10.3390/ijerph16224347.

94 Lisa M. Christian et al., "Inflammatory Responses to Trivalent Influenza Virus Vaccine among Pregnant Women," *Vaccine* 29, *no.* 48, (2011): 8982–8987, doi:10.1016/j.vaccine.2011.09.039.

95 J.B. Layton et al., "Prenatal Tdap Immunization and Risk of Maternal and Newborn Adverse Events," *Vaccine* 35, no. 33 (2017): 4072–4078, doi:10.1016/j.vaccine.2017.06.071.

96 Ibid.

97 Malini DeSilva et al., "Evaluation of Acute Adverse Events after Covid-19 Vaccination during Pregnancy," *The New England Journal of Medicine* 387, no. 2 (2022): 187–189, doi:10.1056/NEJMc2205276.

98 Elyse O. Kharbanda et al., "Evaluation of the Association of Maternal Pertussis Vaccination with Obstetric Events and Birth Outcomes," *JAMA* 312, no. 18 (2014): 1897–1904, doi:10.1001/jama.2014.14825.

99 J.B. Layton et al., "Prenatal Tdap Immunization and Risk of Maternal and Newborn Adverse Events," *Vaccine* 35, no. 33 (2017): 4072–4078, doi:10.1016/j.vaccine.2017.06.071.

100 Pedro Moro et al., "Enhanced Surveillance of Tetanus Toxoid, Reduced Diphtheria Toxoid, and Acellular Pertussis (Tdap) Vaccines in Pregnancy in the Vaccine Adverse Event Reporting System (VAERS), 2011-2015," *Vaccine* 34, no. 20 (2016): 2349–2353, doi:10.1016/j.vaccine.2016.03.049.

101 "VAERS Data," Vaccine Adverse Event Reporting System (VAERS), updated April 7, 2023, https://vaers.hhs.gov/data.html.

102 Aharon Dick et al., "Safety of Third SARS-CoV-2 Vaccine (Booster Dose) during Pregnancy," *American Journal of Obstetrics & Gynecology MFM 4*, no.4 (2022): 100637, doi:10.1016/j.ajogmf.2022.100637.

103 Ibid.

104 "VAERS Data," Vaccine Adverse Event Reporting System (VAERS), updated April 7, 2023, https://vaers.hhs.gov/data.html.

105 Itai Gat et al., "Covid-19 Vaccination BNT162b2 Temporarily Impairs Semen Concentration and Total Motile Count among Semen Donors," *Andrology* 10, no. 6 (2022): 1016–1022, doi:10.1111/andr.13209.

106 Ibid.

107 Malini DeSilva et al., "Evaluation of Acute Adverse Events after Covid-19 Vaccination during Pregnancy," *The New England Journal of Medicine* 387, no. 2 (2022): 187–189, doi:10.1056/NEJMc2205276.

Afterword

1 "HRSA Data and Statistics," HRSA, June 1, 2023, https://www.hrsa.gov/sites/default/files/hrsa/vicp/vicp-stats.pdf.

2 "Vaccines," US Food and Drug Administration, Feb 8. 2023, https://www.fda.gov/vaccines-blood-biologics/vaccines.

3 "How Vaccines are Developed and Approved for Use," Centers for Disease Control and Prevention, Mar. 30, 2023, https://www.cdc.gov/vaccines/basics/test-approve.html#approving-vaccine.

4   "Development & Approval Process (CBER)," US Food and Drug Administration, May 4, 2023, https://www.fda.gov/vaccines-blood-biologics /development-approval-process-cber.

5   US Food and Drug Administration, *Lipitor:Package Insert* (New York, NY: Parke-Davis., a division of Pfizer Inc., updated Apr. 2019), https://www .accessdata.fda.gov/drugsatfda_docs/label/2019/020702s073lbl.pdf.

6   US Food and Drug Administration, *Enbrel: Package Insert* (Thousand Oaks, CA: Immunex Corporation, marketed by Pfizer Inc. and Amgen Inc., updated Sep. 2011), https://www.accessdata.fda.gov/drugsatfda_docs /label/2012/103795s5503lbl.pdf.

7   US Food and Drug Administration, *Stelara: Package Insert* (Horsham, PA: Janssen Biotech, Inc., Bloomington, IN: Baxter Pharmaceutical Solutions, updated Sep. 2019), https://www.accessdata.fda.gov/drugsatfda_docs/label /2016/761044lbl.pdf.

8   US Food and Drug Administration, *Energix-B: Package Insert* (Research Triangle Park, NC: GlaxoSmithKline, 1989), https://www.fda.gov/media /119403/download.

9   US Food and Drug Administration, *Recombivax HB:Package Insert* (Whitehouse Station, NJ: Merck Sharp & Dohme Corp., a subsidiary of Merck & Co., Inc., updated Dec. 2018), https://www.fda.gov/files /vaccines%2C%20blood%20%26%20biologics/published/package-insert -recombivax-hb.pdf.

10 US Food and Drug Administration, *Ipol: Package Insert* (Swiftwater PA: Sanofi Pasteur Inc., updated May 2022), https://www.fda.gov/files /vaccines%2C%20blood%20%26%20biologics/published/Package-Insert -IPOL.pdf.

11 US Food and Drug Administration, *PedvaxHIB: Package Insert* (West Point, PA: Merck & Co., Inc., 1998), https://www.fda.gov/media/80438/download.

12 US Food and Drug Administration, *Hiberix: Package Insert* (Research Triangle Park, NC: GlaxoSmithKline, updated Apr. 2018), https://www.fda.gov /files/vaccines,%20blood%20&%20biologics/published/Package-Insert— -HIBERIX.pdf.

13 US Food and Drug Administration, *ActHIB: Package Insert* (Swiftwater PA: Sanofi Pasteur Inc., updated Mar. 2022), https://www.fda.gov/media/74395 /download.

14 Ross Lazarus, "Electronic Support for Public Health–Vaccine Adverse Event Reporting System (ESP:VAERS)," *The Agency for Healthcare Research and Quality (AHRQ)*, 2010, https://digital.ahrq.gov/sites/default/files/docs /publication/r18hs017045-lazarus-final-report-2011.pdf.

15 US Congress, House—Energy and Commerce; Ways and Means and Senate—Labor and Human Resources, *National Childhood Vaccine Injury*

Act of 1986, H.R.5546, 99th Cong., Part 1., 1986, H.Rept 99-908, https://www.congress.gov/bill/99th-congress/house-bill/5546.

16 "How to Access Data from the Vaccine Safety Datalink," Centers for Disease Control and Prevention, updated Aug. 31, 2020, https://www.cdc.gov/vaccinesafety/ensuringsafety/monitoring/vsd/index.html.

17 "15 U.S. Code § 3710c—Distribution of Royalties Received by Federal Agencies," Cornell Law School, accessed June 23, 2023, https://tinyurl.com/5ym9p4ck.

18 "Conflicts of Interest in Vaccine Policy Making Majority Staff Report," US House of Representatives: Committee on Government Reform, June 15, 2000, https://childrenshealthdefense.org/wp-content/uploads/conflicts-of-interest-government-reform-2000.pdf.

19 Ibid.

20 "CDC's Ethics Program for Special Government Employees on Federal Advisory Committees," Department of Health and Human Services: Office of Inspector General, Dec. 2009, https://oig.hhs.gov/oei/reports/oei-04-07-00260.pdf.

21 "What is Evidence Based Practice?," University of Arkansas for Medical Sciences, Nov. 17, 2022, https://libguides.uams.edu/c.php?g=673659&p=5114477.

22 "Adverse Effects of Pertussis and Rubella Vaccines: A Report of the Committee to Review the Adverse Consequences of Pertussis and Rubella Vaccines," *Institute of Medicine* (1991): 7, doi:10.17226/1815.

23 Kathleen R. Stratton, Cynthia Johnson Howe, and Richard B. Johnston Jr., "Adverse Events Associated With Childhood Vaccines Other Than Pertussis and Rubella Summary of a Report from the Institute of Medicine," *JAMA* 271, no. 20 (1994): 1602–1605, doi:10.1001/jama.1994.03510440062034.

24 Kathleen Stratton et al., "Adverse Effects of Vaccines: Evidence and Causality," *National Academies Press (US)*, (2011): 19, doi: 10.17226/13164.

25 "The Childhood Immunization Schedule and Safety: Stakeholder Concerns, Scientific Evidence, and Future Studies," *National Academies Press (US)*, Mar. 27, 2013, doi:10.17226/13563.

26 Aviva L. Katz et al., "Informed Consent in Decision-Making in Pediatric Practice," *Pediatrics* 138, no. 2 (2016): e20161485, doi:10.1542/peds.2016-1485.

27 "Instructions for Use: Vaccine Information Statement," Centers for Disease Control and Prevention, updated May 12, 2023, https://www.cdc.gov/vaccines/hcp/vis/about/required-use-instructions.pdf.

28 Ross Lazarus, "Electronic Support for Public Health–Vaccine Adverse Event Reporting System (ESP:VAERS)," *The Agency for Healthcare Research and Quality (AHRQ)*, 2010, https://digital.ahrq.gov/sites/default/files/docs/publication/r18hs017045-lazarus-final-report-2011.pdf.

29 Christina D. Bethell et al., "A National and State Profile of Leading Health Problems and Health Care Quality for US Children: Key Insurance Disparities and Across-State Variations," *Academic Pediatrics* 11, no. 3S (2010): S22–S33, doi:10.1016/j.acap.2010.08.011.

Appendix A

1 Lena H. Sun, "More Than 350 Organizations Write Trump to Endorse Current Vaccines' Safety," *Washington Post*, Feb. 8, 2017, https://www.washingtonpost.com/news/to-your-health/wp/2017/02/08/more-than-350-organizations-write-trump-to-endorse-current-vaccines-safety.
2 MSNBC, "Bill Gates Dishes About President Donald Trump Meetings In Exclusive Video" YouTube, May 17, 2018, https://www.youtube.com/watch?v=dY7byG1YGwg.

Appendix B

1 Robert F. Kennedy Jr. to Dr. Francis Collins (June, 21, 2017), https://childrenshealthdefense.org/email-robert-f-kennedy-jr-dr-francis-collins-nih-director-62117/.
2 Committee on the Assessment of Studies of Health Outcomes Related to the Recommended Childhood Immunization Schedule, Board on Population Health and Public Health Practice and Institute of Medicine, "The Childhood Immunization Schedule and Safety: Stakeholder Concerns, Scientific Evidence, and Future Studies," *National Academies Press (US)*, (2013): 13, doi: 10.17226/13563.
3 Committee on the Assessment of Studies of Health Outcomes Related to the Recommended Childhood Immunization Schedule, Board on Population Health and Public Health Practice and Institute of Medicine, "The Childhood Immunization Schedule and Safety: Stakeholder Concerns, Scientific Evidence, and Future Studies," *National Academies Press (US)*, (2013): 9, doi: 10.17226/13563.
4 Jason M. Glanz et al., "A Population-Based Cohort Study of Undervaccination in 8 Managed-Care Organizations across the United States," JAMA Pediatrics 167, no. 3 (2013): 274–281, doi: 10.1001/jamapediatrics.2013.502.
5 "How to Access Data from the Vaccine Safety Datalink," Centers for Disease Control and Prevention, accessed June 26, 2023, https://www.cdc.gov/vaccinesafety/ensuringsafety/monitoring/vsd/accessing-data.html.

6   "Vaccine Safety Datalink Publications," Centers for Disease Control and Prevention, accessed June 26, 2023, https://www.cdc.gov/vaccinesafety/ensuringsafety/monitoring/vsd/publications.html.

7   Mady Hornig, D. Chian, and W.I. Lipkin, "Neurotoxic Effects of Postnatal Thimerosal Are Mouse Strain Dependent," *Molecular Psychiatry* 9, no. 9 (2004): 833–845, doi:10.1038/sj.mp.4001529.

8   "Autism and Vaccines," Centers for Disease Control and Prevention, accessed on June 26, 2023, https://www.cdc.gov/vaccinesafety/concerns/autism.html.

Appendix C

1   Robert F. Kennedy Jr. to Francis Collins (July, 3, 2017), https://childrenshealthdefense.org/letter-robert-f-kennedy-jr-dr-francis-collins-nih-director/.

2   "CDC's Work on Developmental Disabilities," Centers for Disease Control and Prevention, accessed June 26. 2023, https://tinyurl.com/37rd26za.

3   Christina D. Bethell et al., "A National and State Profile of Leading Health Problems and Health Care Quality for US Children: Key Insurance Disparities and Across-State Variations," *Academic Pediatrics* 11, no. 3S (2011): S2–S33, doi: 10.1016/j.acap.2010.08.011.

4   "Welcome to the CHARGE Study Homepage," UC Davis Medical Center, accessed on June 26, 2023,https://beincharge.ucdavis.edu/.

5   "Welcome to the MARBLES Study Homepage," UC Davis Medical Center, accessed on June 26, 2023, https://marbles.ucdavis.edu/.

6   "Welcome to EARLI," The EARLI Study, accessed on Jun. 26, 2023, http://www.earlistudy.org/.

7   "Research on Autism Spectrum Disorder," Centers for Disease Control and Prevention, accessed June 26, 2023, https://www.cdc.gov/ncbddd/autism/seed.html.

8   "National Children's Study (NCS) Archive," US Department of Health and Human Services, National Institutes of Health, accessed June 26, 2023, https://www.nichd.nih.gov/research/supported/NCS.

9   "National Children's Study (NCS)—1.12 GB," US Department of Health and Human Services, National Institutes of Health, Data and Specimens Hub, accessed on June 26, 2023, https://dash.nichd.nih.gov/Study/228954.

10 "NICHD Director Announces Departure," US Department of Health and Human Services, National Institutes of Health, accessed June 26, 2023, https://www.nichd.nih.gov/newsroom/resources/spotlight/092309-Director-Announcement.

11 "Statement on the National Children's Study," US Department of Health and Human Services, National Institutes of Health, accessed June 26, 2023, https://www.nih.gov/about-nih/who-we-are/nih-director/statements /statement-national-childrens-study.

12 "Statement on the National Children's Study," US Department of Health and Human Services, National Institutes of Health, accessed June 26, 2023, https://www.nih.gov/about-nih/who-we-are/nih-director/statements /statement-national-childrens-study.

13 "Environmental Influences on Child Health Outcomes (ECHO) Program," US Department of Health and Human Services, National Institutes of Health, accessed June 26, 2023, https://www.nih.gov/echo.

14 "NIH Awards More than $150 million for Research on Environmental Influences on Child Health," US Department of Health and Human Services, National Institutes of Health, accessed June 26, 2023, https://www .nih.gov/news-events/news-releases/nih-awards-more-150-million-research -environmental-influences-child-health.

15 "ECHO: Environmental Influences on Child Health Outcomes, National Institutes of Health, accessed June 26, 2023, https://www.nih.gov/sites /default/files/research-training/initiatives/echo/echo.pdf.

Appendix D

1 Francis S. Collins, Lawrence A. Tabak, Carrie D. Wolinetz, Diana W. Bianchi, Linda S. Birnbaum, Anthony S. Fauci, Joshua A. Gordon to Robert F. Kennedy Jr., National Institutes of Health, (Aug. 8, 2017), https: //childrenshealthdefense.org/wp-content/uploads/nih-response-dr-collins -to-robert-f-kennedy-jr-8-8-17.pdf.

Appendix E

1 "§300aa–27. Mandate for safer childhood vaccines," United States Code, accessed July 4, 2023, https://uscode.house.gov/view.xhtml?req=granuleid :USC-prelim-title42-section300aa-27&num=0&edition=prelim.

2 "ICAN v. US Dept. of Health and Human Services," US District Court, Southern District of New York, July 9, 2018, https://childrenshealthdefense .org/wp-content/uploads/rfk-hhs-stipulated-order-july-2018.pdf.

Index

NOTES

NOTES

NOTES

NOTES

NOTES

NOTES

NOTES